Reading About the Environment

Reading About the Environment

An Introductory Guide

PAMELA E. JANSMA

LIBRARIES UNLIMITED, INC.
Englewood, Colorado
1993

For Glen, Kaia, and Eli

LIBRARIES UNLIMITED, INC.
P.O. Box 6633
Englewood, CO 80155-6633

Library of Congress Cataloging-in-Publication Data

Jansma, P. E.
Reading about the environment : an introductory guide / Pamela E. Jansma.
viii, 252 p. 17x25 cm.
Includes bibliographical references and indexes.
ISBN 0-87287-985-2
1. Human ecology--Bibliography. 2. Environmental protection--Bibliography. 3. Environmental policy--Bibliography. I. Title.
Z5861.J36 1992
[GF41]
016.3637--dc20 92-12976
CIP

Contents

Introduction

In the past several years there has been an explosion of information on the environment, a trend that shows no sign of abating. Every day the newspaper and television media bombard us with the details of yet another environmental catastrophe, and conservation organizations send us alarming letters requesting donations. The concerned citizen who wants to act to save the planet, whether a business professional, a student, a homemaker, a government employee, an educator, or a union member, needs more information about environmental problems but may feel overwhelmed by the bewildering array of books and magazine articles at the local library. Which ones would be most useful? Which ones address the topic in which I am most interested? Which are aimed at a general level of scientific understanding? Which books will tell me what I can do to help? The purpose of this annotated bibliography is to answer such questions, allowing the busy layperson to find information quickly. The bibliography contains over 750 entries that are either books or popular magazine articles found in most public libraries. Each entry lists the work, author, publisher, and date of publication. The organization, scope, and bias of each work is discussed briefly. The entry mentions whether tables, figures, graphs, glossaries, reference lists, or summaries are particularly noteworthy. Finally, the overall scientific level of the work is addressed so that the reader will know whether to brush up on college chemistry.

For ease of access, the entries are grouped into 16 chapters. Each chapter begins with a summary essay that outlines the important aspects of the chapter topic. Also described are works that are good, general overviews for the individual who only has time to read one or two books. Three symbols at the end of certain bibliographic citations are used to highlight especially noteworthy entries:

****O** good overview of the chapter topic

****S** scholarly treatment of the subject

****F** excellent factual source

The first three chapters consist of introductory and background works on the topics of general science, the environment, and historical and philosophical views of the human relationship with the natural world. The subsequent 11 chapters concentrate on specific environmental problems: overpopulation; air pollution; acid rain; ozone depletion, global warming, the greenhouse effect, and climate change; water pollution; toxic waste; energy and resource depletion; pesticides, cosmetics, food additives, and other chemicals; radiation; land mismanagement; and noise pollution, visual pollution, and pollution in outer space. In addition to references that are good overviews of the problem, each chapter lists specialized sources that focus on one aspect of the problem. In chapter 6 on acid rain, for example, the work *Ill Winds: Airborne Pollution's Toll on Trees and Crops* by James J. MacKenzie and Mohammed T. El-Ashry, a study of the effects of acid rain on forests and food crops, is cited along with the *Acid Rain Information Book*, 2d Edition, by David V. Bubenick, an excellent source of general information. The last two chapters in the bibliography are more general. Chapter 15 discusses health problems in the environment. Unlike works in the earlier chapters, the citations in this chapter concentrate primarily on health issues and secondarily on the environmental conditions that may cause disease. The final chapter lists sources that offer practical suggestions to help improve the state of the environment. Most people probably will find this section the most useful and rewarding.

While using this reference, the individual must remember that all environmental problems are linked. Information on one topic may turn up in a book on another. Pesticides, for example, are not simply hazardous chemicals washed off fruits and vegetables into sewers and groundwater, but are toxins released into the air where they can interact with substances in the atmosphere and travel for thousands of miles before returning to the ground as acid rain. Similarly, the effects of automobile emissions are not limited to smog in our cities, but extend to ozone depletion in the polar regions. And overpopulation is not merely a crisis of food, but also of resources, development, deforestation, land mismanagement, and urbanization.

This bibliography is a research tool and aid for the nonspecialist. It is not a comprehensive guide to the available literature on the environment for the scientific expert. (The individual who needs such a reference can use either the on-line catalogues in public libraries or *Sourcebook on the Environment: A Guide to the Literature* edited by Kenneth A. Hammond, George Macinko, and Wilma B. Fairchild. The latter, however, appeared in 1978 and predates the onslaught of information of the 1980s.) The emphasis here is on books and articles readily available and understandable to the general public. Works whose audience is primarily the scientific community are not included except when they are the definitive treatments of the subject and are typically found in public libraries. Every effort has been made to choose timely and informative sources. The reader is reminded that additional works will appear after the publication of this book and is urged to keep abreast of the latest information by visiting the new releases shelves of the local library or bookstore.

1

General Science

To understand the complexities surrounding today's environmental issues, the reader must have a basic knowledge of many scientific disciplines. The controversy about global warming, for example, stems from variations in the concentrations of substances such as ozone and carbon dioxide at certain latitudes and in different seasons in the Earth's atmosphere, simplistic assumptions in the computer models for climate simulation, uncertainties in the determination of average past global temperatures, and ignorance of the scale and duration of the Earth's natural climatic cycles. These problems draw upon the fields of atmospheric chemistry, paleoclimatology, mathematics, statistics, oceanography, meteorology, and geology. To comprehend the intricacies of the dangers of radioactive and toxic waste disposal to animal life and to water supplies, one must be familiar with hydrology, nuclear physics, chemistry, soil science, ecology, biology, risk analysis, and demography. It becomes easy to see very quickly the interdisciplinary nature and complexity of environmental studies.

With this in mind, the following list of introductory scientific works was prepared. The purpose is to show the interested individual where to go to find the background material necessary to interpret critically the information, to define scientific terms and concepts, and to make informed decisions. The sources vary from texts written for undergraduate and graduate courses to articles from popular magazines. A few works are included for their spectacular illustrations (note the National Geographic Society's *Living on the Earth*, Peter Francis's and Pat Jones's *Images of Earth*, and Kevin Kelley's *The Home Planet*). The level of knowledge the author expects of the audience is stated in each entry.

1 Behrman, Daniel. **The New World of the Oceans: Men and Oceanography**. Boston: Little, Brown, 1969, 436p.

This book details the history of ocean exploration in the modern era and contains numerous anecdotes about pioneers in the study of the oceans. Chapter 22, "Blight in the Sea," is a lucid discussion of pollution in the world's oceans as of 1969 and historical attitudes toward it. One interesting point concerns the discovery of talc, a major constituent of pesticides, in the air in

Barbados in 1965. The book contains many photographs of oceanographers at work. It is thoroughly enjoyable to read and is recommended to anyone interested in oceanography.

2 Berger, W. H., V. S. Smitaak, and G. Wefer. **Productivity of the Ocean: Present and Past**. New York: John Wiley/Interscience, 1989, 471p.

This is the report of a workshop held in April 1988 on the "Gaia problem"—a hypothesis associated with the Greek goddess Gaia—which theorizes that life on Earth acts collectively to maintain conditions suitable for its own survival. The ocean is seen as essential to this process. It exerts a powerful influence on atmospheric carbon dioxide, a greenhouse gas. The book contains concise reviews of all fields involved in studying the ocean. Topics include fluxes to the seafloor, sea-bed processes, and productivity. The introduction gives an excellent overview of the interdisciplinary nature of the subject.

3 Bradley, R. S. **Quaternary Paleoclimatology: Methods of Paleoclimatic Reconstruction**. Boston: Allen & Unwin, 1985, 472p. ****S**

An elementary introduction to climatology, the book offers detailed discussions of all aspects of climate reconstruction, including dating rocks and ice cores, paleooceanography, and pollen analysis. The volume is well illustrated. Although the emphasis is on the climate of the past, the pitfalls described in the volume apply equally to research of the present climate. It is important to realize the difficulties inherent in climate research to understand the problems fundamental to climate forecasting.

4 Briggs, Asa. "Technology and Economic Development," in *Scientific American*, **Technology and Economic Development**. New York: Alfred A. Knopf, 1963, p. 3-18.

Published as a chapter in a *Scientific American* volume, this essay introduces the "problem of how nations can attain a state of self-sustaining growth." The author traces the division of countries into "developed" and "underdeveloped." Less than one-third of the global population lives in the rich, developed sector. Beginning in 1454 with the fall of Constantinople, the author describes the age of European exploration of the New World, discusses the industrial revolution, and ends with the early twentieth century and the two world wars. The article concludes with a statement on the increasing inequality among nations.

5 Brooks, Harvey, and Raymond Bowerd. "The Assessment of Technology," in *Scientific American*, **Man and the Ecosphere**. San Francisco: W. H. Freeman, 1971, p. 209-217.

The authors represent a large panel of scholars gathered by the National Academy of Sciences to address the optimum use of technology and the minimization of its side effects. The paper first appeared in *Scientific American* in February 1970. The panel warns of the dangers of always preparing for major crises and never planning for problems that are less critical. The discussion on burden-of-proof is illuminating: the panel maintains that harmful effects of technology are justified by the benefits but misuse may persist if dangers are not understood in time. The panel's predictions about technological developments are entertaining and thought provoking.

6 Budyko, M. I., A. B. Ronov, and A. L. Yanshin. **History of the Earth's Atmosphere**. Berlin: Springer-Verlag, 1987, 139p. **S

The text discusses the age, distribution, and chemical composition of sedimentary rocks on the Earth's surface in an attempt to infer the composition of the Earth's atmosphere throughout geologic history. The assumption is that the composition of sedimentary rocks records the original signal of the atmosphere. The focus is on atmospheric oxygen and carbon dioxide. References are both Soviet and Western. Additional chapters describe the modern atmosphere, carbon and oxygen cycles, and modeling methodology. Although the text is technical, it is a good source for the individual who wants background information on the pitfalls and variations endemic to atmospheric research.

7 Carson, Rachel L. **The Sea Around Us**. New York: Oxford University Press, 1951, 230p. **O

The Sea Around Us won the National Book Award for nonfiction in 1951. The book provides a human perspective on the history and description of the oceans that cover 70 percent of the surface of the Earth. Topics include: how the oceans formed; how life began in the early oceans; mineral resources; currents and tides; and the hidden areas and life below the surface of the sea. Although much has been learned about the oceans since 1951, this book remains one of the best introductions for the nonspecialist.

8 Chorley, Richard J. (ed.). **Water, Earth, and Man: A Synthesis of Hydrology, Geomorphology, and Socio-Economic Geography**. London: Methuen, 1969, 588p.

This book provides an introduction to the concepts of hydrology and the human role in interfering with the hydrologic cycle. The author states, for example, that 10 percent of the national wealth of the United States in the late 1960s was tied up in structures designed to alter the hydrologic cycle by collecting, diverting, and storing approximately 25 percent of the surface water. These structures include dams, reservoirs, tanks, sewers, and desalinization plants. The book unfolds much like a traditional text in hydrology, beginning with the hydrologic cycle and continuing into basic concepts and processes such as precipitation and runoff. The text contains good illustrations and can be understood by any interested reader.

9 Cloud, Preston, and Aharon Gibor. "The Oxygen Cycle," in *Scientific American*, **The Biosphere**. San Francisco: W. H. Freeman, 1970, p. 57-68.

The evolution of oxygen in the atmosphere is discussed in this article originally published in *Scientific American* in September 1970. The amount of oxygen in the atmosphere is closely related to the number of plants on Earth: oxygen is one of the products of photosynthesis. The exchange between plants and animals is described extensively. The role of oxygen in the human body also is detailed. Excellent figures illustrate the processes of oxidation and oxidative metabolism. Additional topics are oxygen in the geologic record and the human effect on the oxygen cycle.

10 Cole, Lamont. "The Ecosphere," in *Scientific American*, **Man and the Ecosphere**. San Francisco: W. H. Freeman, 1971, p. 11-16.

This article introduces the concept of the ecosphere—a combination of biosphere and ecosystem. The energy cycles of the ecosphere are discussed and illustrated. The chemical elements required by organisms to sustain life are summarized with special emphasis on the five essential elements: oxygen, carbon, hydrogen, nitrogen, and phosphorous. Figures show the estimated relative abundances of some elements in the Earth, in the atmosphere, and in living matter. Another figure graphs the amount of minerals and organic matter lost to the sea each year. This is a superb introduction to the human place in natural cycles.

11 Compton, Robert R. **Interpreting the Earth**. New York: Harcourt Brace Jovanovich, 1977, 553p. ****O**

This is an introductory geology text that concentrates on natural settings as laboratories. For example, the first chapter introduces water and water-related processes, the second chapter discusses a complete river system, the James River of Virginia, and the third chapter describes a seasonal river, the Rio Grande of the southwestern United States. A summary and list of references conclude each chapter. Also included is a section on additional ideas and sources that provides suggestions for projects. The text has numerous illustrations and the index and glossary are very useful.

12 Cook, Robert C., and Jane Lecht. **People!** Washington, DC: Columbia Books, 1968, 63p.

This introductory text on the study of population discusses population, its growth, the population explosion, the effects on our lives, and methods of controlling the increases. Population problems in the United States are described specifically. Other topics include increasing the world food supply, modern health practices, and measuring the rate of population change. The book contains numerous illustrations and tables. The data, however, are from 1968 and are outdated.

13 Deevey, Edward S., Jr. "Mineral Cycles," in *Scientific American*, **The Biosphere**. San Francisco: W. H. Freeman, 1970, p. 82-92.

This general article written for the educated reader was published initially in *Scientific American* in September 1970. The author discusses phosphorous and sulfur, elements that are essential to the function of living organisms. Other elements introduced include iron, cobalt, calcium, magnesium, and sodium. Most of the article, however, focuses on the interaction of phosphorous and sulfur and their compounds, phosphates and sulfates, with the more common elements of the biosphere—hydrogen, carbon, nitrogen, and oxygen. The cycles are superbly illustrated in numerous figures that accompany the text.

14 Delwiche, C. C. "The Nitrogen Cycle," in *Scientific American*, **The Biosphere**. San Francisco: W. H. Freeman, 1970, p. 69-80.

The nitrogen cycle is discussed in this article originally published in *Scientific American* in September 1970. Nitrogen is 79 percent of the atmosphere but cannot be directly used by most living organisms. It must first be treated by specialized plants and animals. The cycle involves nitrogen-bearing compounds, such as ammonium, nitrate, and nitrite. Topics include: the destruction of nitrogen in the Earth; energy yields of chemical reactions important in the nitrogen cycle; types of nitrogen-fixing organisms; and increasing the amount of nitrogen in fertilizers. The article is clearly written and superbly illustrated.

15 Eagleson, Peter S. **Dynamic Hydrology**. New York: McGraw-Hill, 1970, 462p.

This volume summarizes the principal physical processes that are of interest to the engineering hydrologist. The mathematical approach seeks to present a model that is both physically valid and simple. The book begins with a history of hydrology and the hydrologic cycle and moves on to transport processes in the atmosphere and oceans, the composition of the atmosphere, lakes, and oceans, basic fluid dynamics, climate, precipitation, evaporation, and runoff. The author uses many figures to illustrate qualitative aspects of quantitative discussions. The reader is advised to have some understanding of college mathematics.

16 Ehrlich, Paul R. **The Machinery of Nature**. New York: Simon & Schuster, 1986, 315p. ****O**

This noted population biologist has written a nontechnical book that discusses elementary concepts in ecology and evolutionary biology. The emphasis is on explaining the science necessary to understand the world's environmental predicament. Topics include: ecosystems; population ecology and evolution; behavioral ecology; predation and competition; biogeography; and community ecology. The text is lucid and is an excellent introductory volume that everyone will enjoy.

17 Ehrlich, Paul R., Anne H. Ehrlich, and John P. Holdren. **Human Ecology: Problems and Solutions**. San Francisco: W. H. Freeman, 1973, 304p. ****O**

This is a comprehensive introduction to human ecology. The first part of the book addresses environmental problems, including overpopulation, carrying capacity, pollution, and ecological systems. The second part of the book outlines solutions for which the two main approaches are limiting population and changing human behavior. The book closes with summaries and recommendations. The authors are experts in population dynamics and have written an excellent text.

18 Engel, Leonard. **The Sea**. New York: Time-Life Books, 1968, 190p.

This book is just one of the volumes in the Time-Life Nature Library and is beautifully illustrated, well written, and easy to follow. It is an excellent reference for someone just beginning to learn about the world's oceans. All aspects are covered, including the origin of the sea, currents, waves and tides, and animals of the ocean. The last chapter describing humans and the future of the sea is interesting from a perspective 25 years later. One anecdote details a plan to pull a 10-mile iceberg from Antarctica to Los Angeles to provide a source of fresh water. The trip was estimated to take a year and only half the iceberg would melt during the voyage. The appendix lists important dates in the history of ocean exploration.

19 English, T. Saunders. **Ocean Resources and Public Policy**. Seattle, WA: University of Washington Press, 1973, 184p.

The 11 papers in this volume evolved from a series of seminars on natural resources and public policy given at the University of Washington. Each paper is written by an expert in the field and the topics include: coastlines and continental shelves; the marine food chain and its efficiency; food from the sea; energy and resources from the ocean; and policy planning in the marine sciences. The introductory paper, "Man and the Ocean," defines the term resources and outlines the planning process. The papers contain numerous illustrations.

20 Fischhoff, Baruch, Sarah Lichtenstein, Paul Slovic, Stephen L. Derby, and Ralph L. Keeney. **Acceptable Risk**. Cambridge, England: Cambridge University Press, 1984, 208p.

The idea of risk assessment in determining what technologies are safe is addressed in this book. Examples include nuclear power, recombinant DNA research, food additives, and the DC-10 aircraft. In defining a risk strategy, the authors stress the difficulty of such an analysis because of conflicts in social values, lack of facts, and disagreements among experts. Although not directly relevant to environmental problems, this thought-provoking text is included to illustrate how economics and convenience are weighed against environmental issues.

21 Flawn, Peter Tyrell. **Environmental Geology: Conservation, Land-Use Planning, and Resource Management**. New York: Harper & Row, 1970, 313p.

This introductory text focuses on the relationship between geology and the environment. In addition, it emphasizes conservation, land-use planning, and resource management. Figures are excellent and numerous and the text is concise and clear. The final chapter applies the concepts and principles developed in the book to the specific example of Austin, Texas. The large, color geologic map of Austin and its surroundings is valuable. The author does not assume any prior knowledge so it is a good source for all readers.

22 Francis, Peter, and Pat Jones. **Images of Earth**. Englewood, NJ: Prentice-Hall, 1984, 159p.

This is a breathtakingly beautiful book containing photographs taken of the Earth from space. The images are both natural and false color and most were taken during the Skylab project and on shuttle missions. Simple text explains where each photograph is and what it represents. The images are awe-inspiring and humbling. They reiterate how small and finite our planet really is.

23 French, Herbert E. **Of Rivers and the Sea**. New York: G. P. Putnam's Sons, 1970, 318p. ****O**

This book takes a descriptive approach to the hydrologic cycle and discusses the ocean, the atmosphere, lakes, rivers, and underground water. Many historical anecdotes are sprinkled throughout the text, which makes it very entertaining. This is a perfect source for the individual who has an interest in water and hydrology but does not want to be immersed in the quantitative discourses of most introductory hydrology texts.

24 Gregor, C. Bryan, Robert M. Garrels, Fred T. Mackenzie, and J. Barry Maynard. **Chemical Cycles in the Evolution of the Earth**. New York: John Wiley, 1988, 276p.

This text discusses the cyclicity of the geologic record, an important concept to understand when looking at data on fluctuations of chemical species in the atmosphere on short time scales such as 100 years. The book has five chapters. The first introduces checks and balances in the chemical interactions between continental crust and the ocean and the second summarizes the composition of the present and past atmospheres. Most of the second chapter is dedicated to carbon dioxide levels before the interference of humans. The remaining three chapters are much more geological and concentrate on cycles deep below the Earth's surface.

25 Harvey, H. W. **The Chemistry and Fertility of Sea Waters**. Cambridge, England: Cambridge University Press, 1963, 240p.

This book discusses various aspects of the chemical composition of the flora and fauna within the marine environment. Topics include: changes in composition of sea water from the activity of plants, animals, and bacteria; factors that affect the growth of plants; salinity of sea water; major and minor constituents of sea water; and the dissolved gases in sea water. The chapter on the carbon dioxide system is a useful introduction for someone interested in how the gas, a critical component in the greenhouse effect and global warming, behaves. The text is outdated but is still a good source because, unlike more recent books on the subject, it is not quantitative. A background in chemistry or mathematics is not necessary to understand the concepts.

26 Herring, Peter J., and Malcolm R. Clarke. **Deep Oceans**. New York: Praeger, 1971, 320p.

This is a superbly illustrated volume on the deep sea. Photographs, both color and black-and-white, complement the well-written text. Each chapter is penned by a different scientist, each of whom is an expert in the field about which he or she writes. Chapters cover a variety of topics, including the hardware used in oceanographic work, the constituents of ocean water, and the history of ocean research. The majority of the book, however, addresses animal life in the oceans. This is a thoroughly enjoyable book.

27 Holland, H. D. **The Chemical Evolution of the Atmosphere and the Oceans**. Princeton Series in Geochemistry. Princeton, NJ: Princeton University Press, 1984, 582p.

This is a college-level text that discusses the evolution in geologic time of the atmosphere and the oceans with an emphasis on chemical thermodynamics. A strong background in mathematics is highly desirable. Numerous fundamental data tables and derivations are included that may be of interest to the reader attempting to understand the complexity of the atmosphere and oceans and the difficulty of assessing changes in their chemical composition. The reference list is thorough.

28 Horne, R. A. **Marine Chemistry: The Structure of Water and the Chemistry of the Hydrosphere**. New York: John Wiley/Interscience, 1969, 568p.

The book is a reference for the marine scientist and the advanced undergraduate student. It is technical and best understood by someone with a background in chemistry. The figures, however, are excellent and some concepts are easy to grasp without much previous knowledge. Chapters include: "The Structure of Liquid Water"; "The Thermodynamics of Sea Water"; "Transport Processes in Solution"; and "The Chemical Composition of Sea Water." One topic that is rarely covered, radioactive elements in the sea and atmosphere, is also discussed. The appendices are thorough and have abundant information on the physical properties of and the chemical reactions within the oceans.

29 Hutchinson, G. Evelyn. "The Biosphere," in *Scientific American*, **The Biosphere**. San Francisco: W. H. Freeman, 1970, p. 1-11. ****O**

The article was first published in *Scientific American* in September 1970. The biosphere is defined as that "part of the Earth in which life exists." Although the term is vague, it generally encompasses the oceans and the surface of the Earth. The evolution of the biosphere from 3 billion years ago to the present is described and illustrated in a figure. Photosynthesis is explained in detail as well as the important chemical elements for the function of life. An excellent figure summarizes the major cycles of the biosphere, including evaporation/precipitation, photosynthesis, respiration/transpiration, and mineral fluxes.

30 Idyll, C. P. (ed.). **Exploring the Ocean World: A History of Oceanography**. New York: Thomas Y. Crowell, 1969, 280p.

The book tells the history of oceanography. It begins with a summary of the Great Explorers from the Greeks and Phoenicians through the fifteenth century age of exploration to the modern era. Darwin's voyage on the *Beagle* is discussed in detail. The second half of the book describes the world's oceans. The last chapter concentrates on exploration of the sea. This book has numerous illustrations and is written for the nonspecialist.

31 Kelley, Kevin W. **The Home Planet**. Reading, MA: Addison-Wesley, 1988, 256p.

This is an eloquent, lyrical volume containing photographs of the Earth taken from space. The comments that accompany the images are by many of the early astronauts. Their awe and respect for the fragility and miracle of the blue planet is evident everywhere. The pictures illustrate just how small a part of the universe the Earth occupies.

32 King, Thomson. **Water: Miracle of Nature**. New York: Macmillan, 1953, 238p.

This book discusses all aspects of water, the most important substance necessary for life on Earth. The discussion is in two parts. The first is "Water and the World" in which the three states of water are introduced, the oceans are described, and the evolution of life in water is outlined. The second part is "Water and the Life of Man" in which the struggle to control water is chronicled. The book is an excellent descriptive guide of water in all its manifestations.

33 Krauskopf, Konrad B. **Introduction to Geochemistry, 2d Edition**. New York: McGraw-Hill, 1979, 617p.

Written as a text for a college-level class on the problems of chemistry in geology, this volume covers basic concepts, such as chemical equilibrium, acids and bases, free energy, and complex ideas, such as chemical weathering and oxidation and reduction in sedimentation. The focus is on chemical changes that occur on or in the Earth's crust, the outer skin of our planet. At the end of each chapter is a series of problems designed to review the main points, and the solutions are located in the back of the book. Each chapter also contains a list of references and suggestions for further reading. The book is most useful to someone who has a rudimentary understanding of chemistry.

34 Kuzvart, Milos. **Industrial Minerals and Rocks**. New York: Elsevier, 1984, 454p.

The first half of the book is a technical treatment of the minerals and rocks with industrial uses. The discussion of the relationship between geologic materials and future human culture in the last half of the book is noteworthy. The effect of depletion of current natural resources by various means, including political intervention, are contemplated. Also discussed is the effect of the supply of industrial minerals and rocks on the global economy. It is only of interest to someone with a good geological background.

35 Levine, Norman. **Human Ecology**. North Scituate, MA: Duxbury Press, 1975, 469p.

This book was written for an interdisciplinary course on human ecology taught at the University of Illinois. The theme of the work is that humans function in their environment by means of adaptation. The book is divided into four parts, each of which contains several chapters written by different authors. The last part addresses the human confrontation with the environment. The three chapters in this section discuss feeding the human population, the ecology of disease, and the biological effects of pollution. The glossary is indispensable.

36 Matras, Judah. **Populations and Societies**. Englewood Cliffs, NJ: Prentice-Hall, 1973, 562p.

This is an introductory text on the study of human populations with emphasis on world population history, recent growth trends, and consequences of population growth. The population explosion is defined and discussed. The book is divided into three sections: 1) population and social structure; 2) components of population growth and changing composition; and 3) population growth and social change. Figures are numerous. The bibliography is extensive but outdated. The copy of the 1970 United States census form in the appendix is interesting. The text is not technical and is recommended to anyone interested in the subject.

37 Menard, Scott W., and Elizabeth W. Moen. **Perspectives on Population: An Introduction to Concepts and Issues**. New York: Oxford University Press, 1987, 487p. ****S**

The volume is used as a textbook in undergraduate and graduate courses and is a sourcebook for professionals in population studies. The authors have divided chapters into beginning, intermediate, and advanced levels. The chapters are grouped under seven headings, including population growth and decline in history, population theory and policy, poverty, pollution, resources and development, and population research. If papers are chosen with care, the material is easily understood.

38 National Geographic Society. **Living on the Earth**. Washington, DC: National Geographic Society, 1988, 320p.

This is a superbly illustrated volume that describes the various habitats in which humans live. These include: arid lands; the arctic; islands; forests; grasslands; rivers; highlands; and coasts. Some photographs spotlight the intrusion of modern society into pristine wild land. People who question the cost of environmental destruction should pick up this book and look at the natural beauty that surrounds us.

39 Oppenheimer, Valerie K. **Population**. New York: Foreign Policy Association, 1971, 95p.

The text was written by a research sociologist for use in the classroom. Topics include: world population growth in the past, present, and future; the balance between food and population in the Third World; industrialized societies versus the environment; and family planning. Illustrations are numerous and the treatment is straightforward and intelligent. Questions for discussion are listed in the last chapter entitled "Talking It Over."

40 Penman, H. L. "The Water Cycle," in *Scientific American*, **The Biosphere**. San Francisco: W. H. Freeman, 1970, p. 37-45.

This article first appeared in *Scientific American* in September 1970. Water is the most abundant substance in the biosphere. The water cycle, therefore, is critical to living matter. The basic structure and properties of the water molecule, pH, and pF are described. The important transport mechanisms of water—evaporation, ocean circulation, and river discharge—also are detailed. Several figures accompany the text. The article is recommended to individuals who wish to understand the water cycle and its importance to life to better comprehend the effects of water pollution.

41 Raven, P. H., and G. B. Johnson. **Biology**. St. Louis, MO: Times Mirror/Mosby, 1986, 1321p. ****O**

This is an excellent modern introductory text on biology. All the important concepts are covered from evolution to population dynamics to genetics. The overview of current trends in the rapidly advancing science is especially informative. The book makes a superb reference for the general reader.

42 Riley, J. P., and R. Chester. **Introduction to Marine Chemistry**. London: Academic Press, 1971, 465p.

This is a text that was written for the college-level student with a modest knowledge of chemistry. The oceans are treated as a chemically unified system that includes the hydrosphere, the biosphere, and the geosphere. Topics are: the physical properties of seawater; the physical processes in the oceans; major and minor elements in seawater; dissolved gases and organic compounds in seawater; marine sediments; and biological production in the marine environment. The book is well written and illustrated and the index is thorough.

43 *Scientific American*. **The Dynamic Earth**. Special Issue, September 1983, v. 249, n. 3, 202p.

This is a collection of articles that emphasize the constant energy flux within and on the Earth. Topics are: the Earth's core; mantle; crust; ocean; atmosphere; and biosphere. Each article is written by a distinguished scientist who is an expert in his field. The articles are well illustrated and are excellent introductions for the nonspecialist.

44 *Scientific American*. **Man and the Ecosphere**. San Francisco: W. H. Freeman, 1971, 307p. ****O**

This is a collection of articles originally published in *Scientific American* between 1956 and 1971 that addresses topics in population, resources, and the environment. The articles are grouped into chapters entitled: the ecosphere and preindustrial man; limits rarely perceived; the dimensions of intervention; and on management and buying time. Each chapter is preceded by general commentary. The articles are well written and illustrated.

45 *Scientific American*. **The Ocean**. San Francisco: W. H. Freeman, 1969, 140p.

This book is a collection of articles that first appeared in the September 1969 issue of *Scientific American*. As with all its books, *Scientific American* has ensured that the contributing authors are the foremost scientists in their fields and that the articles are beautifully illustrated and lucid. Topics include: the evolution of the ocean floor; the interplay between the ocean and the atmosphere; the environments of the continental shelves and the deep ocean; the nature of oceanic life; and the effects of human exploration and exploitation of the ocean.

46 Shaw, Elizabeth M. **Hydrology in Practice**. Berkshire, England: Van Nostrand Reinhold, 1983, 569p.

Written as an introductory text for undergraduates in civil engineering, this book is also useful to any student with some quantitative background in water studies. The text begins with a discussion of the hydrologic cycle and continues with a description of hydrological measurements, such as precipitation, evaporation, groundwater, and water quality. The last two parts of the book are concerned primarily with analysis and application. References are included at the end of each chapter. Problems are given at the end of the book. Since hydrology is a mathematical subject, the book becomes very technical, very quickly.

47 Shepard, M. L., J. B. Chaddock, F. H. Cocks, and C. M. Harman. **Introduction to Energy Technology**. Ann Arbor, MI: Ann Arbor Science Publishers, 1976, 300p.

This text was written to accompany an introductory course on the technology of the world's energy supply at Duke University. The authors have tried to keep the mathematics at a minimum, but the general reader who has some quantitative background will benefit the most from the discussion. The text begins with an overview of U.S. energy usage (which may be a little out-of-date) and continues with a summary of energy types, including electric power, nuclear power, solar energy, and fossil fuels. The text is well illustrated and basic concepts such as entropy and the heat engine are explained simply and eloquently. The appendices contain equivalency tables that show, for example, that 1000 kilowatt-hours of electricity is 260.5 pounds of bituminous coal.

48 Skinner, B. J. **Earth Resources, 3d Edition**. Englewood Cliffs, NJ: Prentice-Hall, 1986, 208p.

This introductory text on natural resources was written initially in the late 1960s but has been revised since then. It includes a brief review of metallic and nonmetallic minerals, water and energy resources, renewable and recyclable minerals, and abundance of minerals. The author is a distinguished economic geologist and former president of the Geological Society of America. The book is scholarly, readable, and an excellent reference.

49 Stommel, Henry. **A View of the Sea**. Princeton, NJ: Princeton University Press, 1989, 167p.

A leading oceanographer describes the workings of the ocean using the dialogue between the Chief Scientist of a research cruise and the Chief Engineer of the ship to illustrate the circulation of the ocean. The text is quite detailed, and mathematical expressions are used frequently but not always explained adequately. The emphasis is on constructing a velocity and density structure of the ocean from known wind and buoyancy forcing. Despite its quantitative base, the book communicates the sense of adventure and the problems associated with working at sea. It is a good text for someone with a background in mathematics who is interested in ocean circulation—a topic that is essential to water pollution control.

50 Stumm, Werner, and James T. Morgan. **Aquatic Chemistry**. New York: John Wiley/Interscience, 1970, 583p.

This book is a college-level text that is an introduction to chemical equilibria in natural systems by two of the leaders in the field. Its approach uses chemical thermodynamics extensively and assumes previous knowledge of physical chemistry. The extensive discussion of phosphorous, iron, and manganese includes a case study of iron and manganese in groundwater as the result of organic pollution.

51 Sverdrup, H. U., Martin W. Johnson, and Richard H. Fleming. **The Oceans: Their Physics, Chemistry, and General Biology**. New York: Prentice-Hall, 1942, 1087p. ****F**

This volume is a classic text on the oceans that was published initially in 1942 but has been printed in numerous editions since that time. Topics include:

the physical properties of seawater; the distribution of temperature, salinity, and density; the chemistry of seawater; currents; marine organisms; and marine sedimentation. The treatment is thorough and the text is accompanied by many illustrations. Although recent books contain more up-to-date information, this volume remains an essential reference.

52 Walford, Lionel A. **Living Resources of the Sea; Opportunities for Research and Expansion**. New York: Ronald Press, 1958, 321p.

The book was written by a former chief of the branch of fishery biology of the United States Fish and Wildlife Service. It describes what is and what is not known (as of 1958) about marine life. The justification for the study was the belief that land cultivation does not meet the basic food requirements of the world's population. The book is divided into two parts. The first discusses the marine wilderness and covers topics such as the geography of the oceans and the need for improved fishing boats and equipment. The second part outlines the resources of the sea.

53 Wisler, C. O., and E. F. Brater. **Hydrology**. New York: John Wiley, 1949, 419p.

This is one of the first introductory texts written on hydrology. The authors aimed at a college audience that has no previous experience in the field. The central theme is stream flow, its fluctuations and causes. The book is divided into 10 chapters of which chapters I, VII, and VIII are of the greatest interest to the student of environmental problems. The figures are adequate. The text is slightly mathematical, but the quantitative concepts are explained well, making this book suitable for any interested reader. Because hydrology is a fast-growing science, much of the book is out-of-date.

2

Introduction to Environmental Problems

Instantaneous information and technological breakthroughs permit us to know more about our planet in a shorter amount of time than ever before. Satellite images allow scientists to calculate the number of acres of rainforest lost to agriculture each year. Complex instruments measure the levels of greenhouse gases in the atmosphere. The media report every damaging oil spill, leaking toxic waste disposal site, and polluting industrial giant. The result is an increased awareness among the citizens of the global village of the delicate balance of nature that humans are disrupting. People are ready for change. They know that overpopulation, poverty, social inequality, and ignorance are linked to environmental abuse. Pollution is not a problem that starts with the individual polluter but with general attitudes condoned by societies and lax regulations imposed by governments. It is abetted by overconsumption and waste that tax our natural resources. Many of humankind's social, economic, and environmental problems are interdependent; all actions lead to consequences.

Heightened awareness and increased desire for change have created a deluge of information on the state of the environment, the severity of problems, and the relationships among technology, economics, and pollution. Not since the late 1960s and early 1970s have so many books and articles been published, so many benefit concerts been performed, and so many voices been raised for the environment. The motivated reader can find an introductory source about any aspect of the environment of interest. A good place to begin is with books and articles that discuss the environment and human ecology in general. An individual may wish to analyze whether or not problems have reached the crisis stage. Suitable publications include: *The Environmental Crisis, Opposing Viewpoints* by Julie S. Bach and Lynn Hall; *Our Poisoned Planet: Can We Save It?* by Oliver Trager; *State of the World* by Lester Brown and the Worldwatch Institute; and *The Cassandra Conference: Resources and the Human Predicament* by Paul R. Ehrlich and J. P. Holdren. A shocking article that is particularly powerful because of its accompanying photographs is "East Europe's Dark Dawn" by Jon Thompson in *National Geographic*. Another reader, wanting to know more about the economics of environmental pollution, would turn to Alain Enthoven's and A. Myrick-Freeman's *Pollution, Resources, and the Environment* and Orris Herfindahl's and Allen Kneese's *Quality of the Environment: An Economic Approach to Some Problems in Using Land, Water, and Air*. Other individuals seeking good general overviews would examine *One Earth, One Future* by the National Academy of Sciences, *State of the Ark: An Atlas of Conservation in Action* by Lee Durrell, and *Earth* by Paul R. Ehrlich and Anne H. Ehrlich. The reader who needed facts would thumb through *The Cousteau Almanac*—a vast inventory of life on Earth by the noted ocean explorer Jacques-Yves Cousteau. Michael Allaby's *Dictionary of the Environment* is an indispensable aid to the person overwhelmed by terminology. Additional topics include: the vanishing wilderness; energy and the environment; human ecology; dangerous pollutants; technology, waste, and pollution; the role of the government and legislation; conservation; and the long-term outlook.

The publications in this chapter are all introductory overviews suitable for the general reader. Most appeared recently, but some are works that have held up well from the first period of environmental concern—the late 1960s and early 1970s. This category is among the most rapidly expanding of any in this guide—the reader is advised to keep a careful watch for newly published material.

54 Allaby, Michael. **Dictionary of the Environment, 2d Edition**. New York: Van Nostrand Reinhold, 1983, 529p.

This is a general dictionary that explains terminology from a broad spectrum of environmental sciences. The second edition has been updated from the first published in 1977. Unfortunately, the current renewed interest in environmental issues makes the second edition seem in need of slight revision. This is an excellent aid, however, for readers of the environmental literature.

55 Bach, Julie S., and Lynn Hall (eds.). **The Environmental Crisis, Opposing Viewpoints**. St. Paul, MN: Greenhaven Press, 1986, 263p.

Opposing viewpoints are placed back to back to allow the reader to follow a continuing debate throughout this book. Papers are drawn from magazines, books, and newspapers, in addition to being written expressly for the volume. Six papers (three ideas) are argued in each of six chapters: 1) "Is There an Environmental Crisis?"; 2) "Should Corporations Be Held Responsible for Environmental Disasters?"; 3) "Have Pollution Regulations Improved the Environment?"; 4) "Is Nuclear Power an Acceptable Risk?"; 5) "How Dangerous Are Toxic Wastes?"; and 6) "How Harmful Is Acid Rain?" At the conclusion of each chapter is a "critical thinking skill" designed to improve the reader's comprehension and evaluation of the papers. A list of organizations to contact, a bibliography, and an index are included.

56 Bellini, James. **High Tech Holocaust**. San Francisco: Sierra Club Books, 1986, 255p.

This well-researched book addresses the environmental problems created by our technological progress. All types of industrial pollution are described but pesticides and chemicals are emphasized. Of particular interest are the chapters "Poisoned Earth" and "Staff of Life." The author is an alarmist with the point of view that the time in which to act to change the mess is short.

57 Bresler, Jack B. (ed.). **Human Ecology: Collected Readings**. Reading, MA: Addison-Wesley, 1966, 472p.

This collection of papers reprinted from scientific journals addresses the adaptability of human beings to their environments. The first part of the book describes natural environments, and the second part discusses human-made environmental factors. Examples of the latter include: large populations; nuclear bombs; x-rays; cigarette smoking; automobile exhaust; and the space capsule. The papers introduce the reader to the broad field of ecology. The quality of individual papers varies but overall the book is very good.

58 Brown, Lester. **State of the World**. New York: W. W. Norton, 1986.

This is an annual series published by the Worldwatch Institute. The emphasis is on the important environmental crises and issues of that year. In 1986, for example, the topics discussed include economic deficits, the decommissioning of nuclear power plants, and famines in Africa. This is an excellent source of up-to-date information on environmental problems.

59 Congressional Quarterly. **Environment '90: The Legislative Agenda**. Washington, DC: Congressional Quarterly, 1990, 106p.

This publication covers the major environmental legislation that Congress is debating. In discussing each issue, a general overview of the subject is given. Topics are: the "greening" of Congress; clean air; lobbyists; energy; global warming; fuel economy; agriculture; timber; solid waste; nuclear arms; California; "green" campaigns; recycling; environmental protection; priorities; and a brief history of environmental legislation. The section on priorities summarizes most of the major environmental problems, such as the greenhouse effect, acid rain, air pollution, garbage, and nuclear waste. Numerous figures accompany the text. This is an excellent introduction to the concerns of Congress and the American people, suitable for anyone. Annotated references, notes, and an index are included.

60 Cousteau, Jacques-Yves, and the Staff of the Cousteau Society. **The Cousteau Almanac: An Inventory of Life on Our Water Planet**. Garden City, NY: Doubleday, 1981, 838p.

The book is a massive collection of facts organized into 19 chapters and seven parts. All major environmental problems are tackled, including sustainable agriculture; pesticides; acid rain; ozone depletion; water pollution; overpopulation; animal extinctions; rainforest destruction; and energy. Lists of organizations and profiles of notable people are also highlighted in the text. Fortunately, the table of contents and index are thorough, allowing individuals to find information quickly. Illustrations accompany the text. This is a good source of facts on just about anything.

61 Day, David. **The Environmental Wars: Reports from the Front Lines**. New York: St. Martin's Press, 1989, 310p.

This collection of vignettes about people involved in the daily fight against environmental problems is horrifying, hopeful, and informative. By emphasizing the human drama, the author describes the modern environmental movement in concrete and realistic terms. Ivory poaching, toxic waste dumping, rainforest destruction, illegal whaling, and cactus rustling are among the numerous subjects discussed. A bibliography and an index are included.

62 De Bell, Garrett. **The Environmental Handbook.** New York: Ballantine, 1970, 365p.

All aspects of ecology are discused in this book written for the first Earth Day. Topics include urbanization and pollution in cities to conservation of natural resources. Written in a clear and simple style, the book introduced ecology to a large number of people. Although it is outdated, it is interesting for its historical importance. Many of the discussions are still relevant, as are most of the problems.

63 Detwyler, T. R. **Man's Impact on the Environment.** New York: McGraw-Hill, 1971, 731p.

Especially relevant to the theme of the Earth and human affairs, this book is a collection of papers that discusses the ways in which humans affect the environment. Topics include: atmosphere; climate; water; land; soil; and wildlife. Each paper is written by an expert in the field and most are lucid, scholarly, and easy to follow. A list of references follows each paper and is quite useful although a bit outdated. The collection is not technical and is suitable for anyone interested in the subject.

64 Douglas, William O. **The Three Hundred Year War.** New York: Random House, 1972, 215p.

Written by an Associate Justice of the Supreme Court of the United States, this book is a call for action to prevent the ecological disaster that is threatening the Earth. Justice Douglas covers most of the problems facing the natural world, including air and water pollution, radiation, pesticides, garbage, noise, estuaries, mining, wildlife, wilderness, transportation, and land use. The last chapter is a guide to political activism. Source notes, lists of federal environmental agencies and legislative committees, and an index round out the text. Despite being published in 1972, the book is still relevant today.

65 Durrell, Lee. **State of the Ark: An Atlas of Conservation in Action.** Garden City, NY: Doubleday, 1986, 224p.

This book discusses the Earth as an ark, a self-sustaining biosphere. The first chapter analyzes the impact of humans on air, climate, soil, and water. Other chapters describe threats to the creatures who live on the ark both on global and local scales. The last chapter traces the history of the conservation movement. Solutions and courses of actions are offered in the conclusion to the volume. A list of sources and an index are also included. Although information is not catalogued alphabetically, the book has the look of a desk encyclopedia with abundant photographs and illustrations. This is an excellent introduction for the nonspecialist.

66 Ehrlich, Paul R., and Anne H. Ehrlich. **Earth.** New York: Franklin Watts, 1987, 258p.

The concerned authors have written an eloquent book that summarizes how people's actions have led to the environmentally distressed Earth of the late 1980s. Three broad subjects are discussed in three parts. The first, an island of life, emphasizes the fragility of the planet that humans call home. The second, Earth's predicament, analyzes the environmental health of the Earth and the expected cost on the biosphere. And the third, the human response, describes the reaction of humankind to the news that the Earth is not well. A short annotated bibliography and an index conclude the volume. Numerous photographs accompany the text.

67 Ehrlich, Paul R., and J. P. Holdren (eds.). **The Cassandra Conference: Resources and the Human Predicament**. College Station, TX: Texas A&M University Press, 1987, 344p.

This volume contains essays written by people who predicted the environmental problems of today. Topics include overpopulation, air pollution, water pollution, the overuse of fossil fuels, and the depletion of our natural resources. The essays are fascinating and disturbing to read, all the more so because the messages they brought were largely ignored.

68 Enthoven, Alain C., and A. Myrick-Freeman III. **Pollution, Resources, and the Environment**. New York: W. W. Norton, 1973, 285p.

An interdisciplinary collection of papers on the economics of ecology, this volume is divided into four parts: 1) perspectives on the problem; 2) environmental pollution; 3) resources, humans, and the environment; and 4) the economy versus environmental preservation. The introduction gives a good overview of the contents of each part. The section on environmental pollution is the most interesting. Topics include: air pollution and human health; effluent studies in the Delaware estuary; and clean rhetoric and dirty water.

69 Freudenthal, Ralph I., and Susan Loy Freudenthal. **What You Need to Know to Live with Chemicals**. Green Farms, CT: Hill & Garnett, 1989, 182p.

The authors, an internationally known toxicologist and a teacher of writing, have summarized the important arguments and statistics surrounding chemicals in our environment. Subjects they address include: pesticides in food; radon; nuclear waste; pollutants in drinking water; risk assessment; the disposal of hazardous waste; recycling; the greenhouse effect; and acid rain. Two other topics not often included in descriptions of environmental health hazards—the dangers of health foods and natural toxins—also are discussed. The language is simple and direct. This is a good overview on chemicals in our environment for the general reader.

70 Goudie, Andrew. **The Human Impact on the Natural Environment**. Cambridge, MA: MIT Press, 1986, 338p.

This text was designed for an introductory course in environmental studies at the college level. No previous technical knowledge is assumed. Topics include: the impact of humans on vegetation; soil; animals; water; and on climate and atmosphere. Numerous graphs, figures, tables, and photographs accompany the text. An index and an extensive reference list consisting primarily of journal articles conclude the book.

71 Grand Canyon Symposium. **Environment, Man, Survival**. Salt Lake City, UT: University of Utah Press, 1971, 111p.

The proceedings of a symposium held at the Grand Canyon in 1970 on environment, humans, and survival are collected in this volume. Thirteen speakers from universities, nonprofit organizations, and government gave presentations at the meeting of which seven are included here. The chapter titles are: "Pollution and Political Systems"; "The Population Crisis"; "Man's Part in His Own Biological Future"; "Limits on Food Production"; "Implications of Success and Failure in the Control of Pesticides"; "Use Without Abuse of Our Water Resources"; and "Can We Afford Wilderness?" Each author is an expert in his field. This book is highly recommended for anyone interested in a broad overview of the environment.

72 Hart, John (ed.). **The New Book of California Tomorrow**. Los Altos, CA: William Kaufmann, 1984, 433p.

The volume is a collection of writings on the state of California composed between 1961 and 1983 and compiled by a planning and conservation group called California Tomorrow. Topics are: freeways; the water battle between north and south; conservation; Lake Tahoe; urban blight, congestion and development in Los Angeles; urban planning; the energy crisis; and the agriculture industry. The editor comments on each piece and distinguishes his remarks from the original essay by narrowing the column width on the page. Also included is the text of a pamphlet entitled "California going, going, going...." written in 1962 about uncontrolled population growth. The impact of the pamphlet was compared with that of "Common Sense" before the War of Independence. This is a humorous and entertaining book about a serious problem.

73 Harte, John. **Consider a Spherical Cow: A Course in Environmental Problem Solving**. Los Altos, CA: William Kaufmann, 1985, 283p.

The spherical cow is a reference to the initial guess in a simple model to explain a complex problem. It comes from a joke about a farmer who needs to make his cows produce more milk. After much ruminating, a scientist hired to analyze the problem develops a very involved solution by assuming "a spherical cow," clearly a state that the cow does not resemble very well. Such is the approach, however, that researchers must take when attempting to decipher and solve the world's environmental crises. The book does an admirable job of explaining the various methods used and the pitfalls of each.

74 Herfindahl, Orris C., and Allen V. Kneese. **Quality of the Environment: An Economic Approach to Some Problems in Using Land, Water, and Air**. Washington, DC: Resources for the Future, 1965, 96p.

Two economists attempt to show both noneconomists the validity of economic reasoning in approaching environmental problems and economists an awareness of the noneconomic aspects of the problems. Five problems are cited: water pollution; air pollution; and chemical substances in the environment such as pesticides, urbanization, and development of pristine areas. The last chapter details a research strategy. The problems of transportation outlined in the chapter on the physical environment of urban places is insightful. This volume is well written, easy to follow, and suitable for anyone.

75 The Institute of Ecology. **Man in the Living Environment**. Madison, WI: University of Wisconsin Press, 1972, 288p.

Ecologists from all over the world attended a workshop on Global Ecological Problems held in 1971 by the Institute of Ecology. Four major topics were discussed: (1) "Biogeochemical Cycles of Elements Essential to Protein Production"; (2) "Ecological Constraints on Man's Use of Land"; (3) "Terrestrial Food Webs and Diversity"; and (4) "Man's Impact on the Coastal Zone." The topics were sufficiently broad to allow the introduction of numerous secondary and tertiary problems such as toxic substances in the food chain, pollution in the coastal zone, and population management. The figures and glossary are excellent. This book is recommended as a good overview of the interaction between humans and the environment and of the effects of human activity on global cycles.

76 Jennings, Gary. **The Shrinking Outdoors**. Philadelphia: J. B. Lippincott, 1972, 191p.

The author's tone is angry as he describes the polluted world around him and the declining quality of life. The shrinking outdoors refers to times past when people could move on after despoiling their homes. In the twentieth century, that habit could no longer be supported as the world became more polluted and populated. Chapter titles are "Our One and Only Earth"; "1,000,000 Years of Decay"; "The Cluttered Lands"; "The Vanishing Wildlife"; "The Withering Greenery"; "The Spoiling Waters"; "Too Many People"; and "Too Few Who Care." The book was written with young people in mind and may be a bit simplistic. A list of books and organizations deserving your attention is included. The text is slightly outdated.

77 Johnson, Harry G. **Man and His Environment**. Washington, DC: British-North American Committee, 1973, 33p.

This report of the British-North American Committee presents a balanced account of the relationship between economics and the environment. The emphasis is on economics and indeed, the author is a professor at the London School of Economics and in the Political Science department at the University of Chicago. The report has three sections: (1) "Population Problems and Natural Resources"; (2) "Elementary Economics of the Environment"; and (3) "Economic Growth and the Environment." The last section assesses the quality of life relative to economic growth. This book is nontechnical and recommended to everyone.

78 Massachusetts Institute of Technology. **Man's Impact on the Global Environment; Assessment and Recommendations for Action**. Cambridge, MA: MIT Press, 1970, 319p.

Global environmental problems are defined and discussed by scientists. Suggestions for solutions and immediate actions also are included. The treatment is technical and the writing is full of scientific terminology. Figures and tables of data may be of use to the nonspecialist. Despite the quantitative approach, the subject is handled well and thoroughly. The reference list is particularly interesting and complete, although a bit outdated. The book is suitable for individuals who can wade through scientific jargon.

79 National Academy of Sciences. **One Earth, One Future**. Washington, DC: National Academy of Sciences Press, 1990, 196p.

Printed on recycled paper, this book is a comprehensive introduction to global environmental problems and the role that society must play to find solutions. Subjects include: lessons about global warming from the geologic past; the interactions among the atmosphere, oceans, land, and humans; population; global warming and climate change; rising sea level and coastline destruction; ozone depletion; deforestation; and acid rain. Appendix A lists other recommended readings. This is a scholarly work that stresses the urgency of the problem without being alarmist. The book is necessary reading for all individuals.

80 National Research Council, Commission on Natural Resources, Environmental Studies Board. **Long-Range Environmental Outlook**. Washington, DC: National Academy of Sciences, 1980, 198p.

This volume contains the proceedings of a workshop sponsored by the National Academy of Sciences in November 1979. Each chapter is written by a different author who is an expert in the field discussed. Topics include energy, agriculture, toxic substances, and hazardous facilities. The introductory overview summarizes the material in the rest of the book and provides some general recommendations for alleviating environmental problems. The figures are useful. The review of earlier National Academy of Sciences reports is also valuable. The book is interesting because it represents the opinions of experts in environmental studies who try to remain above political and economic considerations.

81 Tax Foundation, Inc. "Pollution Control: Perspectives on the Government Role," in Woods, Barbara, **Eco-Solutions**. Cambridge, MA: Schenkman, 1972.

Several aspects of the government's role in pollution control are described, including federal and state programs, tax subsidies and grants, amount of money available, and estimates of future costs. The problem of setting standards is thoroughly discussed. The text is accompanied by several tables and one graph and the notes are excellent. This is a good, reasonably objective source for someone interested in the specific costs of environmental pollution and control.

82 Thompson, Jon. "East Europe's Dark Dawn," in *National Geographic*, v. 179, n. 6, June 1991, p. 36-63.

This is an eye-opening article about the pollution in Eastern Europe, the legacy of decades of Communist rule that emphasized "industrialization" and employment of the "wage-earning class" at the expense of everything else. Horrifying images include: Upper Silesia, Poland, where 35 percent of the children have lead poisoning; dead forests in Czechoslovakia from sulfur dioxide emissions; unmasked workers handling asbestos; and air masks on schoolchildren on high pollution days. One happy story describes the clean-up of Hungary's Lake Balaton, now a resort. Excellent photographs accompany the text.

83 Thorndike, Edward H. **Energy and Environment: A Primer for Scientists and Engineers**. Reading, MA: Addison-Wesley, 1976, 286p.

The author aims the discussion at the professional scientist and engineer and, as such, the book is very quantitative and assumes a solid background in mathematics, physics, and chemistry. Nevertheless, any motivated reader will find this a valuable source. The approach is a realistic one. Energy and environmental problems are discussed in the context of economics. Cost-benefit analysis is introduced. Energy-related problems addressed in the text include: thermal pollution; radioactivity; air pollution with emphasis on sulfur dioxide and carbon dioxide; and fossil fuel emissions. Because the book is designed to accompany a college course, questions and problems are at the end of each chapter and the answers are located at the back of the book. A glossary of energy units is very useful.

84 Trager, Oliver. **Our Poisoned Planet: Can We Save It?** New York: Facts on File, 1989, 216p.

A wide variety of newspapers and viewpoints are represented in this collection of editorials on environmental issues taken from American and Canadian newspapers. Each clipping is in the original typeface of the newspaper in which it appeared and is accompanied by the date it was published. The clippings are organized into four parts: 1) "Air and Water Pollution"; 2) "Nuclear Energy"; 3) "Chemicals, Toxic Waste, Pesticides, and Asbestos"; and 4) "Global Issues," which covers topics such as Greenpeace, the garbage barge, recycling, ocean dumping, rainforest destruction, ozone depletion, and oil spills. The entries in each chapter are in a loose chronological order. This unusual format makes for very interesting reading. An index expedites finding editorials about specific places, issues, or incidents.

85 Tver, David F. **Dictionary of Dangerous Pollutants, Ecology, and Environment**. New York: Industrial Press, 1981, 347p.

This dictionary includes abbreviations, acronyms, and terminology from the following fields: noise, air, and water pollution; nuclear, solar, and geothermal energy; ecology; coal; hazardous waste; recycling; solar design and construction; biomass; and climatology. It is an excellent reference for the individual who encounters much jargon in his perusal of the environmental literature.

86 Wagner, Richard H. **Environment and Man**. New York: W. W. Norton, 1978, 591p.

The author wrote this book to accompany an introductory course on the impact humans have on the natural world. He assumes no background in the sciences but states that the book would be most beneficial to the college student. The book is divided into seven parts. The first introduces the Earth's physical environments. The last outlines solutions to environmental dilemmas. The middle five discuss the various types and origins of environmental problems, including agriculture and land pollution, extinction, urbanization, radiation, and air pollution. Lists of further reading follow each chapter. The appendix has a compilation of journals that frequently contain articles on the environment.

3

The Human Relationship to the Environment

Throughout history, humans have pondered their place in the natural world. Does the human intellect place humans above all else in the environment? Do their technological achievements allow them to manipulate nature freely without consideration of the consequences for other living creatures? How an individual or society answers these questions characterizes an approach to global environmental problems and dictates an environmental ethic. A positive response suggests development should proceed with little regard to pollution and with no regulatory interference. A negative one, however, supports responsible action and acceptance of guidelines to ensure the well-being of the natural world. Proponents of either extreme may employ dubious methods to achieve their goals. For example, monkey-wrenching, a technique used by radical environmentalists to protect wilderness lands, is sometimes less than legal (note Edward Abbey's *The Monkey Wrench Gang*).

Arguments have been formulated on all sides of the man-versus-nature debate and secondary issues have been raised, such as the concepts of the ecosphere—the part of the Earth that hosts life—and of Gaia—a philosophy that states the Earth as a whole is a living entity that will heal itself (note Barry Commoner's *The Closing Circle* and Michael Allaby's *A Guide to Gaia*). An additional corollary is that technology is viewed as either the solution to all problems or the bane of civilization (note Ian Barbour's *Earth Might Be Fair: Reflections on Ethics, Religion, and Ecology*).

The sources in this chapter address environmental ethics and discuss the ways human's think about and interact with their environment. Some works concentrate on the beliefs of the past (note Frank Bergen, John Black, Clarence Glacken, George Marsh, and Keith Thomas), whereas others focus on the ecology of early civilizations (note Lester Bilsky). Other topics include: environmental policy (note Seyom Brown and others and Lynton Caldwell); religious perspectives on man's relationship with nature (note Frederick Elder); ecological balance (note Barry Commoner); climatic controls on human development (note Ellsworth Huntington); the "doomsday syndrome"—the conviction that man's actions spell his end (note John Maddox); modern radical environmentalism (note Christopher Manes); communal resources (note Bonnie McCay and James McAcheson); aesthetics and the environment (note I. L. McHarg); ecological ethics (note H. J. McCloskey and Holmes Rolston); and challenges to the traditional Western world view (note Keith Thomas). The list is not all encompassing, but merely a sampling of opinions that will give the reader an idea of the intellectual and philosophical treatments that exist.

87 Abbey, Edward. **The Monkey Wrench Gang**. Salt Lake City, UT: Dream Garden Press, 1985 (Tenth Anniversary Edition), 356p.

The noted environmentalist has written several books but this is his most provocative. The experiences of four people who defend the canyon country of the American southwest from corporate and state development are chronicled in this work of fiction. The extreme methods used (frequently against the law) were what gave the gang its name: The technique was termed *monkey-wrenching*. The descriptions of the scarring of once pristine land are emotional and alarming.

88 Allaby, Michael. **A Guide to Gaia: A Survey of the New Science of Our Living Earth**. New York: E. P. Dutton, 1990, 192p.

This is an overview of the scientific hypothesis called Gaia, which holds that bacteria are primarily in control of the future of the Earth. Proponents argue that living organisms, particularly algae and bacteria, regulate matter such as gases and minerals, allowing life to flourish. The author presents the argument for Gaia in a way that the layperson can understand. Unfortunately, no historical background of the scientific debate over the Gaia hypothesis is included. This is an interesting, provocative book.

89 Arthur, Don R. **Man and His Environment**. New York: American Elsevier, 1969, 218p.

The author begins with a discussion of whether humans are animals and concludes that they are. This book also addresses several aspects of humans and their environment. The first part outlines the origins of humanity and the second describes the influence of the environment on our species' ability to survive. The third and fourth parts are the most relevant. The third part documents the effect of humans on the environment, including industrial pollution, polluting effluents, and pesticides. The fourth part suggests compromises for survival and concentrates primarily on population growth. The illustrations are excellent. The text is easy to follow, enjoyable to read, and suitable for anyone.

90 Ashby, Eric. **Reconciling Man with His Environment**. Stanford, CA: Stanford University Press, 1978, 104p.

This book was written by a biologist for the Leon Sloss Junior Memorial Lectures in the humanities. It is an essay that addresses the human attitude to nature in Western societies over the last 100 years and argues convincingly that it has changed little. The slight difference, however, is for the better. The author states that we do not have an environmental crisis—crises are temporary—but a climacteric. The volume is divided into three sections that discuss different aspects of the chain reaction between disclosure of an environmental hazard and political action to control the hazard. The first section defines the chain reaction, the second provides objective parts of the reaction, and the third describes the subjective components of the chain reaction. The volume is enjoyable to read.

91 Barbour, Ian G. **Earth Might Be Fair: Reflections on Ethics, Religion, and Ecology**. Englewood Cliffs, NJ: Prentice-Hall, 1972, 168p.

This is a scholarly and thought-provoking collection of chapters, each written by a different person, that seeks to define a new environmental ethic. The authors reject the two extreme views: 1) Technology will solve everything; and 2) Technology is evil and must be abandoned. Four problems are analyzed (control of pollution, restraint in consumption, redirection of technology, and limitation of population) in the context of the new ethic, which draws on biblical themes, such as stewardship, and modern scientific ideas, such as the interdependence of living things.

92 Bender, David L., and Bruno Leone (eds.). **The Ecology Controversy**. St. Paul, MN: Greenhaven Press, 1981, 148p.

Part of the Opposing Viewpoints series, this book presents arguments written by various authors on both sides of important ecological issues. Two or three debates are included in each of four chapters: 1) Is the world overpopulated?; 2) Do environmentalists help or hinder national interests?; 3) Do environmental problems demand a change in values?; and 4) How much pollution regulation is acceptable? The book also lists ecological organizations and business magazines. A section on discussion activities is designed to improve the skills of the reader.

93 Bennett, John W. **The Ecological Transition: Cultural Anthropology and Human Adaptation**. New York: Pergamon Press, 1976, 378p.

Written by a professor of anthropology at Washington University in St. Louis who specializes in agrarian ecology and communal economics, this book addresses people's relationship with the natural environment. Topics include: images of humans and nature; culture, ecology, and social policy; biological concepts; and adaptation and human behavior. The text is scholarly. The reference list is very thorough, incorporating books in addition to articles in journals. The work will most likely be of interest to people with backgrounds in anthropology.

94 Bergen, Frank (ed.). **The Wilderness Reader**. New York: New American Library, 1980, 372p.

This is an anthology of essays that emphasize the relationship between humans and wilderness. Several important historical figures are included: William Byrd; George Catlin; John James Audubon; and John Muir. The reader who is interested in the musings of philosophers on the environment will find this an invaluable reference.

95 Bilsky, Lester J. **Historical Ecology: Essays on Environment and Social Change**. Port Washington, NY: Kennikat Press, 1980, 195p.

This collection of papers describes environmental problems of the past. The papers are organized into four categories. The first provides an overview of and defines the concepts of historical ecology. The second discusses the ecology of the ancient world, focusing on China and the early Greeks and Romans. The third addresses the ecology of the European Middle Ages. The fourth section outlines the ecology of the modern Western world. The index and footnotes are thorough and the volume is recommended to everyone.

96 Black, John. **The Dominion of Man**. Edinburgh: University Press, 1970, 169p.

The material was originally presented as a series of lectures on conservation entitled "Western World-View and the Inevitability of the Ecological Crisis" in 1969 in London. It gives the historical perspective and outlines the assumptions inherent to the Western world-view of nature and people's relationship to it. The discussion of stewardship and dominion over nature is very valuable. It stresses the change in people's attitudes that has led to accepting responsibility for our power over nature. The volume includes an excellent bibliography and index. This is recommended to anyone interested in environmental philosophy.

97 Brown, Harrison. **The Challenge of Man's Future: An Inquiry Concerning the Condition of Man During the Years That Lie Ahead**. New York: Viking Press, 1954, 290p.

The subtitle introduces the theme of this work. The author examines the human past and present to gather clues for predicting man's future, emphasizing the interaction between people, the natural environment, and technology. Chapters discuss the status quo, vital statistics of the Earth, food, energy, and patterns of the future. The tone of the book is fundamentally optimistic because the author believes people have the ability to extricate themselves from serious predicaments, provided they can recognize and understand the problems. Written in 1953, the book is still timely.

98 Brown, Seyom, Nina W. Cornell, Larry L. Fabian, and Edith Brown Weiss. **Regimes for the Ocean, Outer Space, and Weather**. Washington, DC: Brookings Institution, 1977, 257p.

This book was the result of the Technology and International Institutions Project of the Brookings Foreign Policy Studies Program conducted from 1972 to 1976. The study concentrates on the management of three areas that are outside the jurisdiction of any nation—the ocean, outer space, and the weather. Questions such as "What criteria determine how the three areas are used?" and "Who should decide the rules and who should implement them?" are posed and answered in the context of U.S. foreign policy. In the discussion of these issues, much information is provided on hazards and pollution in the oceans and in outer space, and their effects on climate and the weather. The book is fascinating reading for anyone interested in the subject.

99 Caldwell, Lynton K. **International Environmental Policy: Emergence and Dimensions**. Durham, NC: Duke University Press, 1984, 488p.

This is a guide to the history of international environmental activities. Modern (as of 1984) policy and action also are reviewed. Because of the recent interest in the environment around the world and of increased government cooperation, the volume may be outdated.

100 Cassidy, Vincent H. **The Sea Around Them**. Baton Rouge, LA: Louisiana State University Press, 1968, 202p.

The state of knowledge of the North Atlantic in A.D. 1250 in Western Europe is described in this book. This includes awareness of tides, currents, salinity, islands, and bordering lands. Although not directly relevant to environmental problems, the book is interesting because of the insight it provides into expectations about the sea that were governed by limited knowledge. In modern times with our increased awareness, expectations are greater and the tendency toward exploitation is much stronger.

101 Commoner, Barry. **The Closing Circle**. New York: Alfred A. Knopf, 1971, 326p.

This is a classic volume written by one of the principal spokespersons for the environment. The gloomy picture portrayed by the book motivated people on both sides of the environmental debate in the early 1970s. The author attempts to define the environmental crisis by explaining the ecosphere throughout global history and by noting the changes to its delicate balances. The last part of the book seeks to determine why these balances are broken and finds answers in technology and the economic, social, and political forces that drive technology. The volume is entertaining, insightful, and readable.

102 Crosby, Alfred W. **Ecological Imperialism: The Biological Expansion of Europe, 900-1900**. Cambridge, England: Cambridge University Press, 1986, 368p.

This is a treatise on the migrations of Europeans from the Crusades to shortly before World War I. Other people of the world have restricted their movements to the latitudes in which they originally flourished. Europeans have moved everywhere primarily through their technological means. The author argues that productivity is what has driven the migrations. For example, the United States produces a small fraction of the world's rice but accounts for 20 percent of rice exports. The study is an interesting one that may shed light on the proclivity of industrialized Western nations to pollute. The tone of the book, however, is somewhat elitist and xenophobic.

103 Dubos, Rene. **The Wooing of the Earth**. New York: Charles Scribner's Sons, 1980, 183p.

This biologist and Pulitzer Prize winning author looks at people's relationship with nature through history. A point that is argued convincingly is that animals, including humans, change their environment to suit their needs. The author acknowledges the destruction that humans are responsible for but emphasizes the "environmental enhancements" that they have done, such as the bare slopes of Greece. He is hopeful for the future because of nature's resilience. An appendix lists what the author believes to be the most serious problems facing humankind. An index and a short section of notes conclude the book.

104 Elder, Frederick. **Crisis in Eden**. Nashville, TN: Abingdon Press, 1970, 172p.

This is a religious study of people and the environment. The author outlines an alternative to the traditional Western view of human dominion over nature and alienation from nature. The view is inclusionist with a Christian perspective. Inclusionist is defined as believing the whole of life is greater than the sum of its parts, that life is a unity. The bibliography includes books, articles, and government documents that focus primarily on environmental problems. Some religious sources also are cited. This is an interesting treatment of the environment.

105 English, Paul Ward, and Robert C. Mayfield. **Man, Space, and Environment**. London: Oxford University Press, 1972, 623p.

Geography focused on several themes in the 1960s, including cultural landscape, ecology, environmental perception and behavior, and spatial order. This book addresses each of the major themes in both a historical-cultural and quantitative-economic context. The section on ecological perspectives gives interesting background for understanding and evaluating the reactions of humans to their environment and the development of philosophies of control. A description of the game theoretic framework of humans against the environment is particularly enlightening. The text is suitable for any college-level student.

106 Glacken, Clarence J. **Traces on the Rhodian Shore**. Berkeley, CA: University of California Press, 1967, 763p.

This volume is a thorough and eloquent treatment of the relationship between nature and culture in Western thought from the ancient Greeks to the beginnings of the nineteenth century. The evolution of thought from the planet being designed for humans alone to the planet being created for all living creatures is traced. Until the end of the eighteenth century, the world view was that people bring order to nature, fulfilling their mission from God. The bibliography and index are exhaustive. This book is recommended to anyone with a keen interest in the historical development of the philosophy of the human relationship to the environment.

107 Goldsmith, Edward, Robert Allen, Michael Allaby, John Davoll, and Sam Lawrence. **Blueprint for Survival**. Boston: Houghton Mifflin, 1972, 189p.

Written in the early 1970s by editors of *The Ecologist*, this book outlines a new philosophy for humans' cooperation with the natural world. The authors start from the premise that if present trends continue, the disruption of the life support systems of the Earth is inevitable possibly by the year 2000 and, if not then, definitely by the middle of the twenty-first century. The book has three parts: 1) "The Program"; 2) "The Evidence"; and 3) "The Goal." Population growth is perceived as a major problem. This is suitable for any level student of ecology.

107a Gore, Albert. **Earth in the Balance: Ecology and the Human Spirit**. Boston: Houghton Mifflin, 1992, 407p.

The former senator from Tennessee and now vice-president of the United States has written a comprehensive analysis of the current state of the global environment. The emphasis is on the "balance" between the individual and society and between the environment and economics. The work has an underlying philosophy based on the necessity for feeling "connected to nature." The importance of a worldwide approach to solutions of the serious problems is emphasized. Several topics are discussed, including ecology, overpopulation, technology, and education. Gore, a former journalist, has written a thorough and highly readable book. Notes for each chapter, a good bibliography, and a thorough index conclude the work.

108 Greenwood, Ned H. **Human Environments and Natural Systems: A Conflict of Dominion**. North Scituate, MA: Duxbury Press, 1973, 429p.

The conflict between the environments people create and the natural environments they effect is examined in this book. Ecological balance is described in detail and the options needed to maintain the balance are outlined.

109 Hardin, Garrett. **Filters Against Folly**. New York: Viking Press, 1985, 240p.

The author of the pioneering essay *The Tragedy of the Commons* has written an eloquent and inspiring manual on how to survive the current onslaught of information in our ecologically battered world. He argues that three intellectual tools are needed to act as "filters against folly": 1) literacy or the ability to understand what the words mean; 2) numeracy or the capacity to grasp quantitative information; and 3) ecolacy or the talent to see the consequences of actions over a long period of time. By developing these skills, the author argues that most conflicts between economists and ecologists will be resolved and the solutions will be apparent.

110 Hughes, J. D. **Ecology in Ancient Civilizations**. Albuquerque, NM: University of New Mexico Press, 1975, 181p.

The author traces the attitudes of early civilizations toward the environment. Some of the groups studied include the ancient Greeks, Romans, Egyptians, and Samarians. The text is eloquent and well researched. The book is recommended to individuals interested in the history of environmental philosophy.

111 Huntington, Ellsworth. **Civilization and Climate**. New Haven, CT: Yale University Press, 1924, 453p.

An early work in modern geography, this volume confines itself to the effect of climate on human development and civilization. The primary focus was the importance of physical environment in determining the distribution of civilization. The relation between climate and health also is discussed. Additional topics include: the effect of the seasons, work, and weather; the ideal climate; the conditions of civilization; and the shifting climatic zones. This is a good background source that illustrates the importance of climate in people's adaptability and development, something to keep in mind when pondering global warming.

112 Kimble, George H. T. **The Way of the World**, The Rushton Lecture Series, 1952. New York: George Grady Press, 1953, 123p.

This is a collection of lectures given in 1952 by George Kimble, a notable geographer and former director of the American Geographical Society. The subjects are varied: "The Strength of the Earth"; "The Perseverance of Humans"; and "The Resilience of Society." "The Strength of the Earth" discusses the draw of natural resources on people and their desire to exploit them. The effects of climate, food, and water on the proliferation and activities of humans also are described. Although no particular environmental problems are addressed, the book provides insight into humans' tenuous position on Earth and the ease with which they can alter the conditions necessary for their survival.

113 Lash, Jonathon, Katherine Gillman, and David Sheridan. **A Season of Spoils**. New York: Pantheon Books, 1984, 385p.

The authors look at the environmental philosophy of the Reagan administration during the president's first term in office. The policies of Anne Gorsuch Burford and James Watt, the heads of the Environmental Protection Agency (EPA) and the Department of the Interior, respectively, during those years, are discussed in detail. Despite the drastic increase in environmental pollution and the decrease in federal regulation of industry, the EPA did not expand during the Reagan years. This is a tragic and shocking account of the decline of the monitoring system designed to safeguard our environment.

114 Livingston, John A. **One Cosmic Instant: Man's Fleeting Supremacy**. Boston: Houghton Mifflin, 1973, 243p.

The brief amount of time that people have spent on Earth is emphasized in this book. The planet is 4.5 billion years old. Humans arrived only a few million years ago. Despite this humbling fact, they have always placed themselves above the natural world that sustains them. The author traces this philosophy from classical times through the Christian era to the present. He argues that this belief in absolute authority over the nonhuman world must change if the environment is to be saved. This is a compelling book that demonstrates convincingly the need for a new ethic. A list of references and an index are included.

115 Lorenz, Konrad. **Civilized Man's Eight Deadly Sins**. New York: Harcourt Brace Jovanovich, 1973, 107p.

The author discusses eight separate but related processes that he believes will destroy civilization and threaten humankind as a species. These are overpopulation; destruction of our natural environment; the technological race; overindulgence; "genetic decay"; break in tradition; increased indoctrination; and nuclear weapons. Genetic decay refers to the breakdown of factors that preserve "norms of social behavior." Break in tradition is defined as the younger generation's inability to communicate with their parents' generation. The author is quite reactionary. The chapters that discuss the more obvious environmental problems, such as overpopulation and devastation of natural resources, are the least pedantic. This book is not technical and is suitable for anyone with a penchant for rhetoric.

116 Maddox, John. **The Doomsday Syndrome**. New York: McGraw-Hill, 1972, 293p.

This book was written in response to the flurry of works that were published in the 1960s after the appearance of *Silent Spring* by Rachel Carson, which spelled the end of the world because of people's interference with nature. The author terms this "the doomsday syndrome" and attempts to illustrate why the problems of overpopulation and pollution are not as calamitous as earlier works suggest. He argues that the earlier works ignore the ways in which "social institutions and human aspirations can conspire to solve the most daunting problems." The book is enlightening. The controversial opinions of the author make the work extremely interesting and the reference list is excellent.

117 Manes, Christopher. **Green Rage: Radical Environmentalism and the Unmaking of Civilization**. Boston: Little, Brown, 1990, 291p.

The book is a philosophical treatise that defends the idea associated with groups such as Earth First!, that no action is wrong when protecting the Earth. Deep ecology is discussed in detail. The ramifications of radical environmentalism throughout the world are considered. This is an interesting, well-written, and thoroughly researched book.

118 Marsh, George Perkins. **Man and Nature**. Cambridge, MA: Harvard University Press, 1965, 472p.

Man and Nature was written in the late nineteenth century to illustrate that humans, by tampering with the environment, were making the world uninhabitable. The book was one of the pillars of the early conservation movement in the United States, appearing at a time when Americans believed their natural resources were infinite. The book contains six chapters. The first chapter is an introduction. The three middle chapters are entitled "The Woods," "The Waters," and "The Sands." The last chapter describes geographical changes made by humans. The approach is historical, detailing modifications to the environment since the days of ancient Egypt. The book is recommended to anyone interested in conservation in general and in the early days of the conservation movement in the United States in particular.

119 McCay, Bonnie J., and James M. McAcheson (eds.). **The Question of the Commons: The Culture and Ecology of Communal Resources**. Tucson, AZ: University of Arizona Press, 1987, 493p.

This is a collection of papers written by different authors. The title refers to the theory of the "tragedy of the commons" coined by Garret Hardin, which states that all resources owned in common—air, oceans, wildlife—are, or will be, overexploited. The argument is that normal market forces will not regulate use of the common resources. The papers in this volume address the dilemmas of the commons. Emphasis is on conservation. The bibliography is excellent. The work is not technical.

120 McCloskey, H. J. **Ecological Ethics and Politics**. Totowa, NJ: Rowman and Littlefield, 1983, 167p.

The book begins with a discussion of ecology as a science and its relevance to moral and political concerns. The second part emphasizes ecological ethics and introduces topics such as the preservation of nature, animal rights, and resource depletion. The third part focuses on environmental politics and addresses the politics of preservation and animal rights, political controls of science and technology, pollution, overpopulation, and the politics of ecological reform. The bibliography is excellent. Citations include both environmental and philosophical texts. The writing is dense, but readable.

121 McHarg, I. L. **Design with Nature**. New York: Natural History Press, 1969, 197p.

A bit removed from the main emphasis of the bibliography, this book is included because it is an excellent representative of the topic of aesthetics and its relationship to the environment. Aesthetics and other human values are applied to the question of how to live harmoniously with nature. The greater metropolitan areas of Philadelphia, Baltimore, and Washington and the regions of Staten Island, New York, and the Potomac River Basin are discussed as specific examples. This is a very interesting book and is recommended to all readers.

122 Mines, Samuel. **The Last Days of Mankind**. New York: Simon & Schuster, 1971, 319p.

The author's premise is that anything you pick up is attached to everything else in the universe, a phrase attributed to John Muir. Disturbing anything, therefore, is at your own risk. He discusses urbanization and pollution in cities, people's relationship with the animal kingdom, destruction of waterways, and deforestation. Examples are the Everglades; the Aswan Dam in Egypt; the SST; and public lands in the United States. The bias is in favor of strong conservation efforts. The incidents described are outdated, but the message is still valid. A good index is included.

123 Myers, Norman. **Gaia: An Atlas of Planet Management**. Garden City, NY: Anchor/Doubleday, 1984, 272p.

The author has written an eloquent introduction to the concept of Gaia, the belief that the Earth is a living, breathing organism that sustains and rejuvenates itself. The way the Earth works is summarized. A section on how humankind should manage, not control, the Earth is included. This is a thought-provoking book that everyone will enjoy reading.

124 National Research Council. **The Earth and Human Affairs**. San Francisco: Canfield Press, 1972, 142p.

This is a report prepared by the Committee on Geological Sciences, Division of Earth Sciences in the National Research Council of the National Academy of Sciences. The report tries to outline ways to preserve modern society by using resources most effectively. Five chapters discuss various topics, including the cost of the good life, the Earth through time, people's effect on the land, misuse of land, and cooperation with the Earth in the harvesting of its resources. Radioactive wastes, pollution, soils, and minerals are described in subdivisions of the major chapters. The book is well illustrated. This volume is recommended to anyone interested in the subject.

125 Owen, D. F. **Man's Environmental Predicament**. London: Oxford University Press, 1973, 214p.

This is an introduction to human ecology in tropical Africa. Topics include the relationship between the environment and the people, population, life and death in rural and urban Africa, cultivation, diseases, food and nutrition, natural selection, and development. The author attempts to address the biological consequences of development proposals and urges that solutions must come from within Africa, not from interested third parties. An elementary knowledge of genetics is recommended for the chapter on natural selection and genetics. The bibliography is very thorough.

126 Paehlke, Robert C. **Environmentalism and the Future of Progressive Politics**. New Haven, CT: Yale University Press, 1989, 325p.

The author shows convincingly that environmentalism has implications for political ideologies and personal beliefs. He views modern environmentalism as a progressive alternative to the neoconservatism of the 1980s. The book traces the history of environmental philosophy from the founding of the conservation movement of the late 1800s through the increased concern toward pollution, energy, and natural resources of the 1960s to the heightened awareness of the Earth's habitats in the 1980s. The discussion also emphasizes environmentalism as a system of values and compares it with liberalism, socialism, and conservatism. The last part of the book speculates about the politics of the future.

127 Passmore, John. **Man's Responsibility for Nature**. New York: Charles Scribner's Sons, 1974, 213p.

The first part of this book details Western traditions that either promote or prevent the destruction of the environment by people. The second part outlines the four main ecological problems: pollution; conservation; overpopulation; and depletion of natural resources. The third part discusses how Western attitudes need to change in order for life to flourish on Earth. The author is not scientific in his approach. He uses moral, metaphysical, and political arguments. The book has a good index.

128 Proceedings of the International Congress on the Human Environment. **Science for Better Environment**. Oxford, England: Pergamon Press, 1977, 992p.

This volume contains the proceedings of the International Congress on the Human Environment held in Kyoto, Japan, in November 1975. The lengthy tome has many interesting papers by a variety of authors on topics such as current environmental problems in Japan, toward a better human environment, people's impact on terrestrial ecosystems, and problems in the fields of natural sciences, social science, and cultural studies. The two special lectures on the human place in natural ecosystems and the relationship between environment and technology are noteworthy. Figures are numerous. This volume is recommended to all student levels.

129 Rifkin, Jeremy. **Entropy: A New World View** . New York: Viking Press, 1980, 305p.

Entropy is presented as a law that defines the broad context of our physical world and places limits on human action in that world. The author asserts that the universe began with structure and value and is changing toward random chaos and waste, as dictated by the second law of thermodynamics. He takes this a step farther by stating that entropy law destroys the ideas of history as progress and of science and technology as creating a more ordered world. In chapters 4 and 5, environmental issues are introduced, including the energy crisis, synfuels, nuclear fission and fusion, recycling, conservation, agriculture, urbanization, and health. The book is thought-provoking and is suitable for high-school level students.

130 Rolston, Holmes III. **Environmental Ethics**. Philadelphia: Temple University Press, 1988, 391p.

The author attempts to describe an ethic for the Earth and its "communities of life." He begins with an introduction to how humans value the natural world and continues with an overview of how humans developed dominion over sentient animals and ecosystems. Additional topics include: the application of environmental ethics and value theory to choices in society, government, industry, and business; theoretical views of nature; and historic accounts of people living in harmony with the Earth. Notes are detailed and the bibliography is short but strong. This is a thorough treatment of the topic.

131 Roth, Charles E. **The Most Dangerous Animal in the World**. Reading, MA: Addison-Wesley, 1971, 127p.

The book traces human development from the apes to modern man. In the beginning, people were benign and concerned simply with eating, sleeping, and surviving. As their dominance over the other animals grew and their needs increased, people became the "most dangerous animal in the world." This essay uncovers many of the reasons behind people's relationship with their environment and their lack of concern about its destruction.

132 Russel, W. M. S. **Man, Nature, and History: Controlling the Environment**. Garden City, NY: Natural History Press, 1969, 252p.

This is a superbly illustrated and simply written history of people's desire to control their natural environment. Three ideas, or themes, are followed: human independence from nature by virtue of discovery and invention; the human position as the most important factor on Earth because of the ability to exert pressure on nature; and the human ability to produce surplus food, allowing others to pursue nonagricultural activities. The last required humans to change the surface of the Earth. It is this process on which the book focuses. The author clearly shows environmental problems are not just the result of twentieth century urbanization and industrialization. This book can be enjoyed by anyone. It is easy to follow and it is also a great reference for junior high school students.

133 Sagan, Dorian. **Biospheres**. New York: McGraw-Hill, 1990, 198p.

The author expands on the Gaia hypothesis by arguing that the Earth is in the process of recreating itself. This process sees humans as the "genitals" of the Earth—humankind is engaged in an act of reproduction by constructing self-sustaining ecological enclosures such as the Biosphere II in Arizona. These biospheres may be the only solution to preserving life on a polluted planet, according to the author, and, thus, are the logical manifestations of an Earth trying to save itself. This follows the Gaia hypothesis that the Earth and life on it are an organism that acts to maintain the conditions necessary for its own survival. This is a fascinating discussion that is largely devoid of scientific facts.

134 Sale, Kirkpatrick. **Dwellers in the Land: The Bioregional Vision**. San Francisco: Sierra Club Books, 1985, 217p.

The author, a founding member of the North American Bioregional Congress, explains "bioregional philosophy," a world view that advocates divisions of the Earth's surface based on natural, not political borders. Areas would be defined by their flora, fauna, and geomorphology. The belief has its inception in the concept of Gaia, which holds that the Earth is a living, breathing organism that constantly is rejuvenating itself. Bioregionalism would alleviate environmental crises by ensuring that natural relationships and ecosystems are preserved. This is an interesting discourse on a subject that also is part of the Green Movement. A good index and bibliography accompany the text.

135 Schneider, Stephen H., and Lynne Morton. **The Primordial Bond: Exploring Connections Between Man and Nature through the Humanities and Sciences**. New York: Plenum Press, 1981, 324p.

Chapters address humanistic expressions of cycles in nature, reactions to people's relationship to nature, their place in the world, the philosophy of the ancient Greeks, the nature of science, and the global cycles of life. The chapter on human intervention on global cycles is valuable. The bibliography and index are good. References are included that may not be encountered normally in a study on environmental problems. This is a scholarly and readable interdisciplinary approach that is free of jargon.

136 *Scientific American*. **Man and the Ecosphere**. San Francisco: W. H. Freeman, 1971, 307p.

This text is a collection of 27 articles that first appeared in the magazine, *Scientific American*, between 1956 and 1971. It is an excellent overview of people's interaction with the environment. Articles are grouped into four categories, each of which has its own introduction. The four sections are: 1) the ecosphere and preindustrial man; 2) limits rarely perceived; 3) the dimensions of intervention; and 4) on management and buying time. The third section is the most interesting, covering topics such as air pollution and public health, ionizing radiation and the citizen, and thermal pollution and aquatic life. As in all *Scientific American* collections, the articles are extremely well written and illustrated. This is recommended enthusiastically to anyone interested in the subject.

137 Shepard, Paul. **The Tender Carnivore and the Sacred Game**. New York: Charles Scribner's Sons, 1973, 302p.

This examination of the behavior of humans in the natural world throughout history was written by a noted naturalist. The author asserts that modern man began with the development of agriculture—the first control over nature. The majority of the book describes humans in an adolescent state, the state we were in as hunter-gatherers prior to the dawn of agriculture. The last chapter addresses agricultural man and his removal from nature. The bibliography and index are good. Very artistic drawings highlight the text.

138 Strong, Maurice F. (ed.). **Who Speaks for Earth?** New York: W. W. Norton, 1973, 173p.

This volume contains a series of lectures delivered at the United Nations Conference on the Human Environment in 1972. Each lecture, given by a different expert and presented here as a separate chapter, focused on a major environmental issue. Speakers included: Barbara Ward, a professor of economic development; Thor Heyerdahl, an anthropologist and explorer; and Lord Zuckerman, a former chief scientific advisor to the British government. The topics are the vulnerability of the ocean; the economics of an improved environment; population; and science, technology, and environmental management. This book is entertaining and easy to read.

139 Taylor, Paul W. **Respect for Nature**. Princeton, NJ: Princeton University Press, 1986, 329p.

A theory of environmental ethics, defined as the "moral relations that hold between humans and the natural world" is outlined in this book. Discussion begins with an introduction of environmental ethics and valid moral principles. Subsequent topics include: the attitude of respect for nature; the biocentric outlook on nature; the ethical system; the rights of plants and animals; and competing claims and priority principles. The bibliography is an excellent source for additional readings on environmental ethics and philosophy. This is scholarly and readable.

140 Thomas, Keith. **Man and the Natural World: A History of the Modern Sensibility**. New York: Pantheon Books, 1983, 426p.

This volume addresses changes in people's ideas about their relationship to the natural world that occurred between 1500 and 1800. Thoughts concerning this relationship that flourished in late nineteenth-century England also are discussed. During this period, the human right to exploit other species was challenged. As humans continued to dominate the animal world, their views of that world have shifted away from divine dominion toward coexistence and harmony. The notes for each chapter are extensive and composed primarily of literary sources.

141 Tuan, Yi-Fu. **Topophilia: A Study of Environmental Perception, Attitudes, and Values**. Englewood Cliffs, NJ: Prentice-Hall, 1974, 260p.

The author discusses human views on the physical environment, both natural and human-made. He attempts to answer questions such as: "How do economy and lifestyle affect environmental attitudes?"; "What are our environmental ideals?"; and "What are the links between environment and a world view?" Although the book fundamentally is a philosophical treatise and does not address pollution and ecology directly, it is still of interest to the student of the environmental crisis. The book yields insight into the thought processes that determine human behavior in a variety of good and bad environmental settings.

142 Vogt, William. **Road to Survival**. New York: William Sloane, 1948, 355p.

The book has the thesis that real wealth is drawn from the Earth and is always limited in quantity. It comes from minerals, food, water, and wildlife. Destruction of the Earth's surface and wasting of its products, therefore, deplete the wealth or resource capital of a nation. The author, writing during World War II, saw a severe depletion of resource capital and attempted to focus world attention by describing humans in relationship to their environment. The book is one of the first studies to do so. All concepts are addressed that are relevant to people's interaction with the environment, including carrying capacity of the Earth, birth rates, vanishing waters, industrialization, and the food chain. The text, somewhat outdated, is of historical interest.

143 Ward, Barbara, and Rene Dubos. **Only One Earth: The Care and Maintenance of a Small Planet**. New York: W. W. Norton, 1972, 225p.

The work is an unofficial report commissioned by the secretary-general of the United Nations Conference on the Human Environment held in Stockholm in 1972 in which 58 countries participated. The book begins with a brief history of early civilizations and technical achievements. The introduction provides an interesting perspective by outlining the different points of view of members of the conference. Science, modern technology, land abuse, natural resources, population, and pollution are discussed in later chapters. The book closes with strategies for the future. Figures are limited to bar graphs. This is suitable for anyone who wants an overview of global ecology.

144 Watson, Richard A., and Patty Jo Watson. **Man and Nature: An Anthropological Essay in Human Ecology**. New York: Harcourt, Brace and World, 1969, 172p.

Eight ways of life are described hypothetically. These are the nonhuman primate, the protohuman, and the six stages of humans. Each stage is characterized by an increase in control over the physical environment. The last stage, for example, is the atomic stage whereas an earlier stage is the advanced hunter and gatherer. This book illustrates the anthropologist's view of human interaction with the environment and the escalating desire to dominate the natural world. The perspective is informative—human's basic goals seem to lead inevitably to pollution.

145 Weisberg, Barry. **Beyond Repair: The Ecology of Capitalism**. Boston: Beacon Press, 1971, 201p.

This is a treatise on the destruction of the environment that capitalist societies perpetuate. The emphasis is on the evils perpetrated on the world by the United States such as the bombing of Hiroshima and Nagasaki and the malnutrition of much of the planet brought on by economic colonization. A new structure of the world based on the interdependence of all organisms is outlined, where capitalism, particularly that in the United States, is blamed for ecological imbalances. The book is provocative and entertaining.

146 Whisenhunt, Donald W. **The Environment and the American Experience**. Port Washington, NY: Kennikat Press, 1974, 136p.

This is a historian's view of the ecological crisis in the United States. The author argues that the philosophical, religious, political, and economic systems all have contributed to the present state of the environment. The situation is the result of views held by the Western world for hundreds of years. The book begins with an introduction to the problem. Subsequent chapters focus on the intellectual, religious, economic, political, and population revolutions, and the structure of the future. The book has a good index. This is enjoyable reading for the individual who does not mind pedantic writing.

147 Wilson, E. O. **Biophilia**. Cambridge, MA: Harvard University Press, 1984, 157p.

The author, a noted evolutionist and ecologist, has written an eloquent and concise philosophical work on the idea that humankind has a natural empathy for other living things. The book is uplifting and hopeful. The fundamental belief, however, is difficult to accept at times when viewing the horrible acts that occur daily on our planet. Everyone should listen to this optimistic voice in the darkness.

148 Worster, Donald (ed.).**American Environmentalism, the Formative Years, 1860-1915**. New York: John Wiley, 1973, 234p.

This is a collection of classic papers written between 1860 and 1915 by noted naturalists such as John Wesley Powell who first navigated the Colorado River and John Muir who explored the Sierra Nevada of California. Papers are grouped into four parts: 1) "Man and Nature in an Age of Science"; 2) "The Conservation Movement"; 3) "Garden, City, and Suburb"; and 4) "The Biocentric Revolution." The suggestions for further reading are primarily scholarly historical studies written in the 1950s and 1960s.

149 Young, John. **Sustaining the Earth**. Cambridge, MA: Harvard University Press, 1990, 225p.

The author, a professor of history and the director of the Center for Environmental Studies at the University of Adelaide in Australia, has written a timely assessment of the global environmental movement. The work has eight chapters that discuss the need for an improved environment, the role of science, the required common ground, the ingredients of post environmentalism, and the politics of a sustainable society. The movements and strategies of the last few decades are summarized. This is a refreshingly pragmatic and realistic view of the problem and its possible solutions. Bibliography, references, and an index are included. Suggestions for further reading are at the conclusion of each chapter.

4

Overpopulation

In 1985, leaders from nations that contain over half the world's population signed a Statement on Population Stabilization that said, "Degradation of the world's environment, income inequality, and the potential for conflict exist today because of over-consumption and over-population. If this unprecedented population growth continues, future generations of children will not have adequate food, housing, medical care, education, earth resources, and employment opportunities." The global population is 5.2 billion. The United Nations predicts this figure will increase to 10 billion by 2025. This stands in sharp contrast to earlier in human history when doubling rates were on the order of hundreds of years. In short, the number of people who are alive today is greater than the number of people who have ever lived. Adding severity to the problem is that most of the population growth is in underdeveloped nations where environmental destruction, in the form of deforestation, desertification, soil erosion, and natural resource depletion, is extreme.

Cries of alarm at the rapid population growth rate were sounded in the mid-1900s. (Earlier in the century, scientists and philosophers argued whether overpopulation or underpopulation was the significant issue—note works by Enid Charles, T. H. Marshall and others, and Pitt-Rivers). A pamphlet distributed in the early 1950s introduced the terms *population bomb* and *population explosion* to the American people (this event is described by Lawrence Lader in *Breeding Ourselves to Death*). Another warning was given in *The Prevalence of People* by Marston Bates in 1955. He argued that by eradicating once fatal diseases, modern medicine contributed to death by starvation in the world's most populous countries. Perhaps the most influential treatment of the problem was *The Population Bomb* written by Paul Ehrlich, a noted biologist, in 1968. Through this work, the term "carrying capacity"—the maximum number of individuals that can be supported in a given habitat as limited by natural resources, food, and so on—became familiar to the public. The author also outlined the consequences that numbers of people close to the carrying capacity of the Earth might have on the environment, including increased air and water pollution, greater demand for natural resources and food, and additional toxic waste. This message was amplified further in *The Population Explosion* written by Paul R. Ehrlich and

Anne H. Ehrlich in 1990, which states that overpopulation is the most urgent environmental problem. Global warming, rainforest destruction, ozone depletion, pollution, and famine arise from the surplus of people and the resultant drain on the Earth's natural resources. A similar point of view is held by Garrett Hardin in *Exploring New Ethics for Survival: The Voyage of the Spaceship Beagle* published in 1972. He turns more philosophical, however, debating whether parenthood is a right or a privilege among other issues.

The works in this chapter discuss a variety of topics related to global population. These include: population growth and development; human population and food production; population and natural resources; limits on global population; reasons for population growth; population and agriculture; and demography and population theory. Most of the books and articles were written in the 1960s and 1970s when overpopulation was seen as the most pressing global environmental problem. The list serves as a sampling of the more accessible treatments and as an overview of the broad topic of overpopulation. Many more sources exist than are cited here. The reader who seeks additional information is advised to peruse the bibliographies of the relevant citations. Global population statistics can be found in reports published by federal governments, the United Nations, and the National Academy of Sciences.

150 American Assembly. **The Population Dilemma**. Englewood Cliffs, NJ: Prentice-Hall, 1963, 188p. ****F**

This is an overview of population problems in which each chapter is written by a different author. Key facts about population, problems created by increasing growth, policy issues, and action alternatives are discussed with an emphasis toward providing the reader with the background necessary to make informed decisions about population studies. Chapter titles include: "World Population Growth"; Population and Economic Development"; "Population Growth in the United States and the World"; and "Issues in Population Policy." The book was prepared as a guide for the American Assembly and contains their final report. Some figures accompany the text.

151 Appleman, Philip. **The Silent Explosion**. Boston: Beacon Press, 1965, 161p.

Written by a professor who spent months traveling through the most overpopulated regions of the world, this book attempts to alert the United States and other technologically advanced societies to the population explosion. The moral, economic, and social implications of population growth and control are discussed. The ideologies of Marxism and Catholicism are presented in the context of population growth. The chapter "Our Country and Theirs" stresses the importance of industrialized and scientifically knowledgeable countries to undertake research in the effects of uncontrolled population growth and to offer possible solutions. This is well written and easy to follow. It also has a good index and many notes.

152 Bahr, Howard M., Bruce A. Chadwick, and Darwin L. Thomas. **Population, Resources, and the Future**. Provo, UT: Brigham Young University Press, 1972, 352p.

This is a collection of papers that focus on the idea that population is not a crisis but rather the failure of society to provide for the increased numbers of people. The most critical factor is believed to be per capita energy consumption, not population growth. Resources also are considered adequate and no shortfalls are projected by the authors. The perspectives are "non-Malthusian", i.e., the writers do not subscribe to the view that population growth is out of control and must be limited.

153 Bates, Marston. **The Prevalence of People**. New York: Charles Scribner's Sons, 1955, 268p. **S

The book was written by a biologist at the request of a health specialist to ponder the question whether public health and medical research were alleviating or creating problems on a global scale. The argument was that by ensuring survival through once fatal diseases, modern medicine was contributing to death by starvation in the world's most crowded and undeveloped countries. What resulted is a lucid and eloquent book that defines the terms population problem and overpopulation and describes the processes that govern growth in societies.

154 Bourgeois-Pichat, Jean. **Population Growth and Development**. New York: International Conciliation, Carnegie Endowment for International Peace, 1966, 81p.

This study of world population growth and its implications draws on information submitted to the United Nations World Population Conference in Belgrade in 1965. The author attempts to discuss the problem without oversimplifying it. Topics include: the demographic distress of the world and the threat to economic development; projections for the year 2000; fertility; mortality; international migration; and the role of international organizations. The text is accompanied by 29 tables and 11 graphs. This is interesting to read over 20 years past its publication date.

155 Brown, Harrison, and Edward Hutchings, Jr. **Are Our Descendants Doomed?** New York: Viking Press, 1972, 377p.

This is a collection of papers by noted experts on technological change and population growth. Topics are varied and include: changing birth and death rates; the relationship among population, GNP, and the environment; population planning and the Catholic church in Latin America; and unemployment. Each paper is followed by a brief discussion by an author with a different point of view. Some disagree violently whereas others simply alert the reader to misstatements and questionable observations. The introduction is a good overview of human social development through the ages from less populous to more populous societies.

156 Brown, Lester R. "Human Food Production as a Process in the Biosphere," in *Scientific American*, **The Biosphere**. San Francisco: W. H. Freeman, 1970, p. 93-103.

The article, first published in *Scientific American* in September 1970, holds that the growth of human population is controlled largely by increases in food production. The author, an expert on agriculture and population, begins with a brief history of agriculture and continues with a discussion of modern fertilization and irrigation practices. The consequences of fertilizers in our atmosphere and water are weighed against their benefits. DDT is cited as an example. Some solutions to the problems of pesticides are suggested. The article is scholarly and superbly illustrated.

157 Brown, Lester R. **In the Human Interest: A Strategy to Stabilize World Population**. New York: W. W. Norton, 1974, 190p. **S

The author insists that the population of the world must be kept under 6 billion and that our lifestyles must be modified drastically. While holding this view, the author presents a strategy to achieve his goals. The scholarly book is divided into five parts: 1) "The Demographic Backdrop"; 2) "Population Growth on a Finite Planet"; 3) "Common Dependence on Scarce Resources"; 4) Population Growth and Social Conditions"; and 5) Confronting the Population Threat." The second and third parts are the most relevant. Subtopics include: endangered species; eutrophication of lakes; soil erosion; overloaded waste capacity; food; energy; and minerals. The suggested reading list is excellent. The book is scholarly and readable.

158 Carr-Saunders, A. M. **World Population, Past Growth and Present Trends**. London: Frank Cass, 1964, 336p.

This is an introduction to the study of world population, or, as the author states, "how men count themselves." All important aspects of population are discussed, including migration, deaths, births, fertility, and population growth. A considerable portion of the middle of the book is devoted to population studies in Europe where population was declining. A note on overpopulation concludes the book. The volume is fairly technical and is recommended to someone with considerable interest in the field.

159 Charles, Enid. **The Twilight of Parenthood**. New York: W. W. Norton, 1934, 226p.

Although this book is dated it presents the point of view that underpopulation, not overpopulation, is the great threat of the time. The author states that advances in food production techniques have changed the relationship between humans and natural resources such that the world population can be handled easily. She also cites new statistical procedures that clarify birth and death rates to allow new deductions to be made concerning population growth. The work is interesting in its historical context.

160 Clark, Colin. **Population Growth and Land Use**. London: Macmillan, 1967, 406p.

This is an introductory work on population growth and its effect on land use. The book is divided into nine chapters, including the reproductive capacity of the human race, survival and growth, population and food, location of industries and population, and land use in urban areas. The author does not have the opinion that rapid population growth leads to poverty but rather that growth forces communities to become more technologically advanced and productive, thereby making themselves more wealthy. The only drawback to population growth the author feels is serious is the rapid spread of cities. There is a good index and numerous graphs, tables, and charts.

161 Dyke, Bennett, and Jean Walters MacCluer (eds.). **Computer Simulation in Human Population Studies**. New York: Academic Press, 1974, 518p.

This collection of papers describes computer models for various aspects of population. Categories include: anthropology and social systems; genetics and adaptive systems; demography; and simulation methodology. The subjects of the papers yield insight into possible pitfalls encountered in population research. Most articles are technical and best understood by someone with either a mathematical background or computer familiarity. The bibliography is very thorough but is dominated by texts on computer simulations. The index is good.

162 Ehrlich, Paul R. **The Population Bomb**. New York: Ballantine Books, 1968, 223p. ****O**

The author is a professor of biology at Stanford University and is a noted expert in the field of population biology. This book is a classic study of the consequences of uncontrolled population growth that has inspired numerous other works. The author describes the effects of overpopulation on our air, food, and water supplies and on our overall environment. The book begins with an overview of the problem and continues with an outline of what is being done and what still needs to be done. It finishes with suggestions for what the individual can do. The bibliography lists many books of general interest. This is essential reading for anyone concerned about overpopulation.

163 Ehrlich, Paul R., and Anne H. Ehrlich. **The Population Explosion**. New York: Simon & Schuster, 1990, 320p. ****O**

A sequel to *The Population Bomb*, this book emphasizes that overpopulation is the number one environmental problem. Global warming, rainforest destruction, ozone depletion, pollution, and famine are secondary. The authors warn that time for action is overdue and suggest possible solutions to control population growth. The opening chapter, intended to fend off criticism, discusses why not everyone believes that population increases are bad. Extensive notes and an index are included.

164 Ehrlich, Paul R., and Anne H. Ehrlich.**Population, Resources, Environment; Issues in Human Ecology**. San Francisco: W. H. Freeman, 1970, 383p. ****O**

This text introduces the reader to relationships among population, resources, and the environment. The authors are noted experts who have written numerous books and articles on the threats of overpopulation. The text contains 13 chapters. Titles include: numbers of people; population structure and projection; optimum population and human biology; and birth control. Three chapters are of particular interest: 1) "The Limits of the Earth"; 2) "Environmental Threats to Man"; and (3) "Ecosystems in Jeopardy." Figures complement the written material nicely. References are at the end of each chapter. This text is highly recommended to anyone interested in the stresses populations exert on the ecosphere.

165 Farmer, Richard N., John D. Long, and George J. Stolnitz (eds.). **World Population—The View Ahead**. Bloomington, IN: Indiana University Press, 1968, 310p.

This volume is a collection of papers by experts in the field of population studies presented at the conference on world population problems held at Indiana University in May 1967. Papers on population growth are grouped under five headings: 1) "The Outlook"; 2) "Economic and Technological Aspects"; 3) "Sociological Aspects and Quality of Life"; 4) "Death and Birth Controls"; and 5) "Ethical Issues of Control." The papers are technical but not quantitative. The introductory overview gives an outline of the main points of the book. This volume is interesting from an historical perspective.

166 Fraser, Dean. **The People Problem: What You Should Know About Growing Population and Vanishing Resources**. Bloomington, IN: Indiana University Press, 1971, 248p.

In this book, the author does not cast blame on anything specific for the population problem but views it as a natural consequence of human behavior and people's desire for life. The author does, however, believe the situation is critical and amasses evidence supporting his contention. Topics include: exponential expansion of the population; factor limiting population growth; and population control. Food, water, minerals, energy, and pollution are subheadings. A list of suggested readings and an index are at the end of the book.

167 Frejka, Tomas. **The Future of Population Growth**. New York: John Wiley, 1973, 268p.

The author states that this book is for "social and economic planners, civil servants, academicians, politicians and interested citizens." Issues addressed include: when to decide a country's or region's population should no longer grow; if growth is attainable; and what the consequences of current growth trends are. The methods and assumptions used in the study are outlined prior to discussion of specific countries and their population problems. The treatment is technical, and, in places, mathematical. A quantitative background is recommended.

168 Glantz, Michael H. "Drought in Africa." *Scientific American*, v. 256, n. 6, June 1987, p. 34-41.

The author illustrates how the policies of governments encourage the devastation of recurrent droughts in sub-Saharan Africa. He argues that drought is part of the climate of the region. Policy planners should accept that when the rains come, drought is not far behind. The article contains an analysis of the possible causes of drought, an estimate of the food production per capita in the region, and a list of options for the future. Well illustrated and written, this is a good introduction to the climate and population problems of sub-Saharan Africa.

169 Grigg, David. **Population Growth and Agrarian Change**. Cambridge, England: Cambridge University Press, 1980, 340p.

An historical perspective of population growth and concomitant changes in agricultural practices, the book contains four parts. The first describes the methodology, defining overpopulation and outlining the symptoms of overpopulation in agrarian societies. The second, third, and fourth parts are titled "Malthus Justified," "Malthus Refuted," and "Malthus Returns," respectively.

Malthus was a pioneer in population studies, first alerting people to the dangers of overpopulation and establishing rules that govern the way societies increase their numbers. "Malthus Returns" describes modern developing countries. The remaining two parts concern Europe of the fourteenth, sixteenth, and seventeenth centuries. The illustrations, notes, and index are thorough.

170 Hardin, Garrett. **Exploring New Ethics for Survival: The Voyage of the Spaceship Beagle**. Baltimore, MD: Penguin Books, 1972, 273p. ****O**

The name of the spaceship comes from the name of the ship in which Charles Darwin made his spectacular voyage in the nineteenth century. The scientist-author concentrates on problems in the Earth's "backyard" by creating a journey on a spaceship and discovers that overpopulation is the most serious. He questions whether parenthood is a right or a privilege, and addresses numerous other philosophical issues concerning population and development, such as private property and agrarian reform. The book is a thought-provoking, well-written treatment of a complex and controversial problem. The essay, "The Tragedy of the Commons," is included in the appendix.

171 Hardin, Garrett. **Population, Evolution, and Birth Control: A Collage of Controversial Readings**. San Francisco: W. H. Freeman, 1964, 341p.

This volume is a collection of papers and essays written over the last few hundred years that addresses certain aspects of population, evolution, and birth control. Included are selections from such notable works as *Utopia* by Thomas More and *Trust in God* by Martin Luther. The pioneer work in population studies, "An Essay on the Principle of Population" by Thomas Robert Malthus is a highlight. Other authors in the volume are Charles Darwin, Matthew Arnold, and Han-Fei Tzu. Modern scholars also are included. The book is intriguing.

172 Hartley, Shirley Foster. **Population Quantity vs. Quality: A Sociological Examination of the Causes and Consequences of the Population Explosion**. Englewood Cliffs, NJ: Prentice-Hall, 1972, 343p.

The book provides an historical perspective on population growth that the author believes is necessary to understand the problems of the world in the late twentieth century. Chapters early in the book discuss the nature of the population explosion whereas later chapters present the consequences of population growth. Topics include the ramifications of increased population on quantity of food and natural resources, economics and standard of living, and family planning. The book has numerous illustrations. It is scholarly and enjoyable to read.

173 Hauser, Philip M., and Otis Dudley Duncan. **The Study of Population**. Chicago: University of Chicago Press, 1959, 864p.

A volume that outlines the status of the science of demography in the United States as of 1959, the book contains numerous papers of interest to readers who want an increased understanding of the methods used by people who study population. The discussion of population studies in several disciplines is particularly enlightening. Papers in this section address a variety of topics, including ecology and demography, human ecology and population studies, genetics and demography, and economics and demography. The book is somewhat technical. The index is thorough.

174 Hinrichs, Nokl (ed.). For the Council on Population and the Environment. **Population, Environment, and People**. New York: McGraw-Hill, 1970, 227p.

The introduction states that this book is for those who know overpopulation is a problem that must be solved and understand the complexities of the issue. The book has four parts. The first addresses the basic problem with a particularly good essay by Paul Ehrlich. The second part discusses the human factors involved in population growth and the concomitant environmental destruction. The third part describes possibilities for action, including stabilizing the United States' population, community action, and environmental education. The final section places the population problem in a larger context by emphasizing the human aspects. This is a good collection of papers by noted experts.

175 Keyfitz, Nathan. **Demography Through Problems**. New York: Springer-Verlag, 1984, 141p.

This is a book in the series of problem books in mathematics. It is an introductory text on population theory and analysis. The format is a series of problems where each is a self-contained puzzle, the solution of which prepares the student to answer the next problem in the sequence. Each problem emphasizes a point about population. Chapters are divided into a first section on mathematical concepts and a second part on problems. Some short computer models are also included. Although the author asserts only "algebraic manipulation" is required, most problems assume familiarity with differential and integral calculus. As such, this is recommended to an individual interested in population studies who has a firm background in calculus.

176 Lader, Lawrence. **Breeding Ourselves to Death**. New York: Ballantine Books, 1971, 120p.

This book outlines the events surrounding the distribution of a pamphlet to the press and leaders of the American government by the Hugh Moore Fund, a nonprofit educational foundation, in the early 1950s. The pamphlet described the problem of overpopulation and, for the first time, alarmed the American people. It coined the terms population bomb and population explosion. The history of the pamphlet is chronicled. In addition, the techniques used by the Hugh Moore Fund are detailed such that other activists may employ the methods to their benefit. Numerous photographs of political figures and newspaper headlines are included.

177 Marshall, T. H., A. M. Carr-Saunders, H. D. Henderson, P. R. Kuczynski, and Arnold Plant. **The Population Problem**. London: Allen & Unwin, 1938, 176p.

This book is drawn from a series of radio talks given in 1937. The first chapter is a summary of the opening debate on what the people of England felt about the population problem. The remaining chapters present the views of the authors on such subjects as the economic consequences of population growth; population trends and international migration; and causes and solutions of the population problem. The treatment is narrow-minded, focusing on Britain and the British empire. Nevertheless, it is an interesting early treatment of a problem that is not perceived as serious because it had not affected British colonial ambitions.

178 McCormack, Arthur. **The Population Explosion**. New York: Harper Colophon Books, 1973, 78p.

This Christian approach to the population explosion was written by a priest who spent many years as a missionary in Africa. The author attempts to address questions of population growth and control that have arisen within the Catholic church. Topics include: food and population; population and urbanization; population and environment; human rights and population; the attitude of the Catholic church; and population and development. The treatment is balanced and optimistic.

179 National Academy of Sciences, Committee on Science and Public Policy. **The Growth of World Population**. Washington, DC: National Academy of Sciences, 1963, 38p.

This is the report of the Panel on Population Problems, National Academy of Sciences that was convened in 1963. The members of the panel express alarm at the doubling rate of the population (every 35 years the population doubles) but are confident that overpopulation can be alleviated by developing methods for "fertility regulation" and implementing family planning programs on a global scale. The summary of world population problems is a good introduction. Also discussed are social and biomedical factors that contribute to population growth.

180 National Academy of Sciences, Office of the Foreign Secretary. **Rapid Population Growth: Consequences and Policy Implications**. Baltimore, MD: Johns Hopkins Press, 1971, 696p.

This volume contains a summary section on the problems of population growth and an additional section composed of papers by various authors. The summary discusses topics such as demographic setting, consequences of rapid growth, and population policy. The overview is excellent. Research papers focus on a variety of issues. Among those debated are the economics of population control, the effects of population growth on resources and food supply, the health consequences of overcrowding, and ethics in the formation of population policy. The index is thorough and references are located at the end of each paper.

181 Osborn, Fairfield (ed.). **Our Crowded Planet: Essays on the Pressures of Population**. Garden City, NY: Doubleday, 1962, 240p.

Edited by a former president of the Conservation Foundation and of the New York Zoological Society, this volume is centered around the belief that overpopulation is the most serious problem in our global society. Papers written by authors from a variety of disciplines address topics such as population pressures on the human natural environment, population pressures on economic and political trends, population and religion, and population pressures on morality and ethics. A selected list of readings is included. Opinions in the volume are diverse and make for interesting reading.

182 Pitt-Rivers, G. H. L. F. (ed.). **Problems of Population**. Port Washington, NY: Kennikat Press, 1932, 378p.

This volume contains the proceedings of the second general assembly of the International Union for the Scientific Investigation of Population Problems held in London in 1931. It is an interesting historical work that addresses underpopulation in addition to overpopulation. Topics include: population and food supply; fertility; population density; population theory; and migration. Papers are written by experts in the field and are variable in quality. Overall, the book is lucid and informative, although it is outdated. Diagrams are numerous. A few papers have abstracts that summarize the main points of the article. The book contains a good index.

183 Population Reference Bureau. **World Population Growth and Responses 1965-1975**. Washington, DC: Population Reference Bureau, 1976, 271p.

The Population Reference Bureau was founded in 1929 to analyze and report changes in population in the world as well as in the United States. The Bureau also assesses the impact of the changes on quality of life and the environment. The bulk of this report contains cut-and-dried information such as birth and death rates for individual countries. The introductory section, however, gives an excellent overview of the global population and programs that have been implemented to halt the increase. World population data for 1965 and 1975 are included. The glossary is short but adequate.

184 Proceedings of the American Society of International Law Regional Meeting and the John Bassett Moore Society of International Law Symposium. **The World Population Crisis: Policy Implications and the Role of Law**. Charlottesville, VA: University of Virginia Press, 1971, 173p.

This is a summary of the proceedings of a meeting held at the University of Virginia Law School in March 1971 that focused on the policy implications and the role of law in the world population crisis. The volume is divided by sessions of which there are four. Selected papers follow the session discussions. Each session begins with several lectures by prominent people in the field of law, policy, and medicine and closes with a debate of the topics presented. Topics include policy goals toward the world population crisis, institutions that may meet the crisis, domestic and international laws, and task force reports. The material is somewhat dry but it is well written.

185 Shelsnyak, M. C. (ed.). **Growth of Population, Consequences and Controls**. New York: Gordon and Breach, 1969, 458p.

The volume contains the transcripts of discussions among experts in the field of population control at the first conference on population held at Princeton, New Jersey, in September 1968. Several topics are debated including "What are the consequences of population growth for the developed and developing countries?" "How good is family planning?" "Can the biological environment be manipulated?" and "Whose business is population growth?" The interesting format of the discussion allows excellent presentation of opposing points of view and construction of arguments. The book is enlightening and entertaining.

186 Singer, S. Fred. **Is There an Optimum Level of Population?** New York: McGraw-Hill, 1971, 426p.

This is a collection of papers by noted scholars in the field of population studies. Topics include: the limitations natural resources and environmental factors place on population; the effectiveness of education, health, and welfare services on a growing population; and the changes in the quality of life as population increases. The discussions on natural resources such as minerals, energy, and food are particularly valuable. The volume has several appendices, the most interesting of which is appendix C that contains messages on population growth from then-President Richard Nixon to various government agencies. The text is scholarly and readable.

187 Smil, Vaclav. "China's Food," in *Scientific American*. v. 253, n. 6, December 1985, p. 116-125.

The article analyzes China's ability to feed its population, one-fifth of the world's total, having only one-fifteenth of the Earth's arable land. Harvests and per capita availability of food have increased in the decades since Mao Tse-tung, allowing for optimism. The use of synthetic fertilizers and the generation of toxic wastes, however, threaten the recycling process that is essential for continued increases and is traditional in ancient Chinese agriculture.

188 Snow, C. P. **The State of Seige**. New York: Charles Scribner's Sons, 1969, 50p.

The book is based on a series of lectures given at Westminster College in November 1968. The author remarks on the state of uneasiness people were experiencing during the few years previous when he was travelling throughout North America, Europe, and the Soviet Union. The uneasiness was brought about by feeling constricted and limited by population growth and the immediacy of technological problems. He closes his discussion with possible solutions for the industrialized countries. This is easy to read and enjoyable.

189 Spengler, Joseph. **Population Change, Modernization, and Welfare**. Englewood Cliffs, NJ: Prentice-Hall, 1974, 182p.

This study emphasizes the global economic aspects of population growth. Data are classified primarily by country. The book has 10 chapters that focus on a variety of topics, including population growth and distribution, population optima, population policy, and population and modernization. The chapter on population and the natural environment is of particular interest. Within this chapter are discussions on the organic environment, sources of organic products, and minerals. A table of arable land in the world is informative. The index is thorough.

190 SPR Charter. **For Unto Us a Child Is Born**. San Francisco: Applegate Books, 1968, 52p.

This is an overview of population growth and its problems based on works from "Man on Earth," a periodical on human ecology, that attempts to define a theory and program for the human place on Earth. The author wrote the book for members of the "well-fed minority" to teach them that they are responsible for increased population pressures and world hunger. The essay is structured around the feelings people have toward the birth of children. Once it was considered a happy occasion; now it is a cause for concern. The discussion is short but interesting.

191 Stanford, Quentin H. (ed.). **The World's Population: Problems of Growth**. Toronto: Oxford University Press, 1972, 346p.

Written for the nonspecialist, this is a collection of 33 articles by experts in the field of population studies. The papers are grouped into three parts. Part I provides the necessary background for understanding the population problem. Terms used in demography are defined. Part II addresses the causes of the population problem, concentrating on mortality and fertility. Implications of the accelerated growth rate also are outlined. Part III presents possible solutions to the population problem. World population statistics for 1970 are included in an appendix. The bibliography and index are adequate. The few figures that accompany the text are valuable.

192 Thompson, Warren S., and David T. Lewis. **Population Problems**. New York: McGraw-Hill, 1965, 593p.

The book presents the facts about population studies to the nonspecialized college student or the lay reader interested in population growth. The statistical methods used are the simplest available. Topics include: the study of population, which has a good discussion of Malthus, one of the pioneers of population studies; the composition of population; major demographic processes such as fertility and birth rates; changing patterns in population growth; and implications of world population changes. The introductory text is well written and illustrated.

193 Wogaman, J. Philip. **The Population Crisis and Moral Responsibility**. Washington, DC: Public Affairs Press, 1973, 340p.

This collection of 34 papers by various authors attempts to clarify the ethical issues raised by the debate on what responses society should have to control global population growth. Papers are grouped into four parts: 1) the moral basis of policy objectives; 2) the moral responsibility of government; 3) moral analysis of policy proposals; and 4) moral responsibility of religious communities. Each part begins with an introduction to and summary of the papers that follow. Although not directly relevant to the problem of overpopulation, this book is included because of the unusual interpretations of population statistics made by some of the authors.

194 Young, Louise B. (ed.). **Population in Perspective**. New York: Oxford University Press, 1968, 460p. ****S**

This is a collection of noted works by experts in the field of population studies. The first paper is the classic work by Thomas R. Malthus, which alerted the world to the overpopulation problem. Papers are divided among four sections: 1) biological factors that influence population such as death rates, food and fertility, and birth control; 2) economic and political factors; 3) religious and social factors; and 4) prospects for the future. Each section has an introduction that explains the themes and summarizes the papers in that section. Terms used regularly by the authors are explained in the foreword. The illustrations and index are good.

5

Air Pollution

The air we breathe is no longer clean. In most urban areas of the United States, air pollution in the form of smoke, soot, smog, and fog exceeds federal standards. The Environmental Protection Agency (EPA) estimated that in 1987 over 2.4 billion pounds of chemicals linked to cancer, genetic defects, and neurological and respiratory diseases were released into the air. The major polluters were the companies that produce chemicals, paper, rubber, plastics, metals, petroleum, and electronic goods. What is particularly frightening about the EPA investigation is that it did not include automobile emissions and industrial factories that spew less than 75,000 pounds of hazardous material into the atmosphere each year.

The most common pollutants regulated by the EPA since the 1970s include: *sulfur dioxide*, a colorless gas formed during the burning of fossil fuels which when combined with water vapor in the atmosphere produces sulfuric acid that is highly corrosive to many things and damaging to the lungs; *nitrogen oxides*, red-brown gases created during the high temperature combustion of fossil fuels that when exposed to sunlight and hydrocarbons produce photochemical oxidants—ozone among them—the primary constituents of smog; *carbon monoxide*, a colorless gas that causes headaches, fatigue, and even death and forms during the burning of fossil fuels when not enough oxygen is present; *particulates*, small dust or liquid droplets produced during the combustion of fossil fuels and suspended by air currents in the atmosphere; *hydrocarbons*, volatile organic compounds, such as benzene, methane, and butane, formed during the incomplete burning of fossil fuels and evaporation of solvents and oil spills; and *lead* emitted from smelting and the making of batteries. Hydrocarbons and lead are among the most toxic air pollutants to humans: the former can cause mutations and cancer whereas the latter can affect internal organs and the nervous system by depriving the blood stream of oxygen.

In addition to causing disease and irritating the respiratory system, air pollution has other undesirable side effects such as acid deposition and crop destruction, depletion of the ozone layer, buildup of greenhouse gases, and global warming. These major issues are covered in chapters 6 and 7.

Pesticides, which when released into the atmosphere produce still another form of air pollution, are discussed in chapter 11.

Perhaps the greatest surprise to most people is that indoor air can be more than five times as hazardous as outdoor air. In contrast to hazy brown smog or orange sunsets, few clues exist to alert one to indoor pollutants because they are frequently odorless and colorless. Among the most dangerous are formaldehyde (an irritant found in particleboard, draperies, and building materials), asbestos (a fibrous substance used in insulation), lead (a metal that may be a constituent of plumbing pipes, paints, and automobile exhaust), and radon (a gas emitted naturally by certain types of rocks and soils). Other pollutants that are detected more easily include tobacco smoke, insect spray, aerosol cleaners, ammonia, and kerosene.

The above description of the most common pollutants and their origin gives the impression that air pollution is a modern problem. It is not. Evidence of poor air quality dates back to the days of ancient Egypt from which blackened mummified lung tissues remain. Also well known is the extreme pollution in London during the Industrial Revolution when coal was burned regularly (note *The Big Smoke: A History of Air Pollution in London Since Medieval Times* by Peter Brimblecombe). Other examples of prehistoric air pollution come from Peru, Alaska, and the Canary Islands.

The subject of air pollution embraces several aspects. Among those included in this chapter are: the coal industry (note Bruce Ackerman and William Hassler, and Elizabeth Haskell); general industrial pollution (note James Cannon); federal regulatory efforts (note the Conservation Foundation entries, Azriel Teller, Lawrence White, and Lester Lave and Gilbert Omenn); smog and automobile emissions (note Susan Faludi and A. J. Haagen-Smit); air pollution and public health (note Walsh McDermott); the latitudinal distribution of carbon monoxide and its relationship to the burning of rainforests (note Reginald Newell, Henry Reichle, Jr., and Wolfgang Seiter); particulates (note Robert Shaw); indoor pollutants (note National Research Council and Assembly of Life Sciences, and Isaac Turiel); tobacco smoke (note Arlene Galloway); and radon (note David Bodansky and others, Michael Lafavore, Mueller Associates and others, and William Nazaroff and Anthony Nero, Jr.). General overviews also are included (note *The Toxic Cloud* by Michael H. Brown, *The Sky Is Still Falling* by Donald E. Carr, *Contemporary World Issues, Environmental Hazards: Air Pollution* by E. Willard Miller and Ruby Miller, "Hazards in the Air" by Eric Skjei and M. Donald Whorton, and "The Air Around Us" by Richard Wagner). Two scholarly and technical approaches are *Air Pollution* by Henry Perkins and *Air Quality* by Thad Godish.

195 Ackerman, Bruce A., and William T. Hassler. **Clean Coal Dirty Air; Or, How the Clean Air Act Became a Multi-Billion Dollar Bail-Out for High-Sulfur Coal Producers and What Should Be Done About It**. New Haven, CT: Yale University Press, 1981, 1983p.

The authors, two legal scholars at Yale University, follow decisions by the Environmental Protection Agency and Congress that led to the regulations that required scrubbers be installed in smokestacks of electric power plants. They argue that despite the appearance of improving air quality, the interest of the American people was not served. The result was the protection of a market for coal from Eastern mines, a product that does not burn cleanly. Although the book is an interesting account, it does not encourage one's faith in the political system.

196 Bodansky, David, Maurice A. Robkin, and David R. Stadler. **Indoor Radon and Its Hazards**. Seattle, WA: University of Washington Press, 1987, 147p.

This is an excellent introduction to indoor radon and its health effects written by a group of professors in physics, nuclear engineering, and environmental health at the University of Washington. Ten chapters discuss a variety of topics, including terminology used in radon investigations, methods for radon detection, sources of radon, normal levels of radon, modification of radon levels in homes, links between radon exposure and lung cancer, and comparison of indoor radon to other radiation hazards. The first chapter is an overview of the radon problem. A glossary and index conclude the book. References are at the end of each chapter. This is an informative treatment that can get technical.

197 Brimblecombe, Peter. **The Big Smoke: A History of Air Pollution in London Since Medieval Times**. London: Metheun & Company, 1987, 185p.

The author begins by introducing the oldest known evidence of the effect of smoke on human beings: the blackened, mummified lung tissues from Egypt, Peru, and the Aleutian and Canary Islands. The account continues with descriptions of modern medical techniques that allow researchers to estimate the effects of smoke on medieval man by examining skulls. The importation of coal to England was a milestone in pollution and its control. The author has done an admirable job of tracing the horrors of air pollution through the ages, proving that it is not just a scourge of modern society and coming to the conclusion that, finally, in present-day London, the problem has abated. Air pollution has decreased 80 percent since the late 1800s and both domestic and industrial sources are under control.

198 Brown, Michael H. **The Toxic Cloud**. New York: Harper & Row, 1987, 307p.

The author describes what is in the air we breathe and where it originates. He discovered that chemical plants in Texas are so large they cause their own weather and increase the incidence of cancer in people living between eastern Texas and western Florida; that air in farm communities is laced with pesticides and herbicides including dioxin; and that people growing up in New Orleans have a one-in-three chance of contracting cancer. The book is well researched and written and the story is alarming.

199 Cannon, James S. **A Clear View; Guide to Industrial Pollution Control**. New York: INFORM, 1975, 246p.

A manual of sorts written for the American citizen who wishes to become involved in the fight to control pollution, this book outlines the problem of industrial pollution. Industry and factory production also are discussed. Chapters on gathering and evaluating factory data are valuable. The remainder of the book details ways in which an individual can have an impact, citing methods such as lobbying and lawsuits. Environmental impact statements are explained. The references are good and include magazines as well as technical books. The glossary and index are thorough. The volume is recommended to anyone thinking of becoming involved.

200 Carr, Donald E. **The Sky Is Still Falling**. New York: W. W. Norton, 1982, 278p.

The author is a former director of research at a major oil company. He has written a compelling book about the dangers of air pollution in our modern society. Problems include: the increase in environmental cancers; small-particle pollution produced by diesel engines; the carcinogens in wood smoke; and coal as an energy source. The last is viewed as particularly nefarious; coal is used as a substitute for nuclear power. The health hazards from the burning of coal, however, are worse than those from the radiation given off by nuclear power plants. Alcohol is proposed as the relatively pollution-free energy resource of the future.

201 The Conservation Foundation. "Toxics in the Air: Reassessing the Regulatory Framework," in The Conservation Foundation, **State of the Government: A View Toward the Nineties**. Washington, DC: The Conservation Foundation, 1987, p. 479-534.

This chapter in a report by the Conservation Foundation about the state of the environment in the mid-1980s emphasizes strategies and risks in the regulatory procedure of air pollution. The discussion begins with descriptions of the risks from airborne toxic substances in the home, office, industry, and motor vehicles. Additional topics are: federal strategies for managing airborne toxic substances, such as the Clean Air Act, the Occupational Safety and Health Act, and the Consumer Product Safety Act; spending at the federal level; state, local, and private sector programs; and identifying risks. Suggestions for further reading conclude the chapter. The reference list is extensive but is dominated by technical journal articles. This is a scholarly and well-researched work.

202 The Conservation Foundation. "Traditional Environmental Contaminants—Air Quality," in The Conservation Foundation, **State of the Government: A View Toward the Nineties**. Washington, DC: The Conservation Foundation, 1987, p. 54-86.

This is a section within the second part of a report designed by the Conservation Foundation to allow informed decisions to be made about environmental policy. Topics include: pollution indicators; the pollution standard index; trends for individual pollutants such as sulfur dioxide, carbon monoxide, and ozone; and long-range atmospheric problems such as the greenhouse effect and acid rain. The text is dominated by statistics and is accompanied by numerous graphs and tables. This is an excellent source of factual information and is recommended to people in search of actual pollution levels.

203 Cowling, Ellis B. **Pollutants in the Air and Acids in the Rain: Influence on Our Natural Environment and a Challenge for Every Industrial Society**. Berkeley, CA: University of California, College of Natural Resources, Department of Forestry and Resource Management, 1985, 26p.

Given initially as a lecture, this essay on the general problems of air pollution focuses on three themes: 1) pollutants and acids in the air; 2) the effect of air pollution on the natural world; and 3) the threat of air pollution to modern society and the solutions most reasonable to combat it. This is a good overview for the general reader.

204 Elsom, Derek. **Atmospheric Pollution: Causes, Effects, and Control Policies**. New York: Basil Blackwell, 1987, 319p.

This is a thorough investigation of air pollution that seeks to answer the questions: What is air pollution? Where does it originate? and What is its effect on the things around us? The destruction of vegetation, agricultural crops, wildlife, and climate are discussed. Also included is an analysis of international cooperation to fight air pollution that looks at the policies of the industrialized nations.

205 Faludi, Susan. "Selling Smog," *California Business*, June 1987, p. 68.

The article discusses the concept of "emission trading," a practice in which companies who have good pollution records are rewarded and those who do not are punished. The author complains that dirty companies have no incentive to comply with regulations because they can always pay penalties and buy "emissions-reductions credits" instead of investing money in control equipment. The loser is the general public who must breathe the smog.

206 Galloway, Arlene. **The Smoke-Free Guide, How to Eliminate Tobacco Smoke from Your Environment**. Victoria, Canada: Qualy, 1988, 159p.

The author, a registered nurse, has written a practical manual on the establishment of smoke-free working areas. Her approach is positive and comprehensive. The book has 10 chapters grouped into four sections: 1) "First the Facts," an introduction to tobacco smoke and its health consequences; 2) "Clearing the Air in Your Personal Environment," a guide to quitting if you do smoke and to maintaining clean air around you if you do not; 3) "Clearing the Air Through Legislation," an overview of state, provincial, and federal laws and ways to implement local ordinances; and 4) "Clearing the Air World-Wide," a discussion of smoking in the Third World. Three appendices, of which the first is a list of statements that the reader must guess are either true or false, conclude the book. Reading this book is an excellent first step in an attempt to remove tobacco smoke from your environment.

207 Gilpin, Alan. **Air Pollution**. St. Lucia, Australia: University of Queensland Press, 1971, 67p.

Although a bit dated, this volume is a good introduction to air pollution for the general reader. All aspects of the problem are addressed. Topics include: the causes and effects of air pollution; major air pollution accidents; clean air legislation; pollution control equipment; and the economics of polluting the air versus keeping it clean.

208 Godish, Thad. **Air Quality**. Chelsea, MI: Lewis Publishers, 1985, 372p.

The emphasis of this book is on air pollution as an interdisciplinary problem. Aspects of chemistry, climatology, biology, physics, meteorology, political science, earth science, and economics are introduced to augment the discussion. Subjects include: pollutants and their consequences; health effects; motor vehicle emissions; indoor air pollution; noise pollution; and regulatory procedures. This is a good source for the general reader who wishes the central issues presented clearly and concisely.

209 Graedel, Thomas E., and Paul J. Crutzen. "The Changing Atmosphere," *Scientific American*, v. 261, n. 9, September 1989, p. 58-68.

Human activity has changed the composition of the Earth's atmosphere faster in the last 200 years than in the rest of human history. The effects of these changes include acid rain, smog, and depletion of ozone in the stratosphere. These effects have occurred largely without altering the concentrations of the major gases, nitrogen and oxygen, which compose greater than 99 percent of the atmosphere. The authors try to answer three questions: 1) Which activities produce what emissions? 2) How do the emissions of trace gases generate the effects? and 3) What are the consequences? The article is very informative and the figures are excellent. Two tables that show the source of trace gases and their contribution to environmental problems are particularly useful.

210 Haagen-Smit, A. J. "Air Conservation," in Bresler, Jack B. (ed.). *Human Ecology, Collected Readings*. Reading, MA: Addison-Wesley, 1966, p. 389-409, also in *Science*, v. 128, 1958, p. 869-878.

The author is the biochemist who first identified smog in Los Angeles as being caused by automobile exhaust, temperature inversions in the Los Angeles basin, and sunlight. The article discusses dust, smoke, Los Angeles smog, automobile exhaust, and cigarette smoke. The section that outlines the effects of Los Angeles smog on plants and rubber is eye-opening. Figures and photographs accompany the text. The references are primarily from scientific journals.

211 Haagen-Smit, A. J. "The Control of Air Pollution," in *Scientific American*, **Cities: Their Origin, Growth and Human Impact**. San Francisco: W. H. Freeman, 1973, p. 150-157.

This article was first published in *Scientific American* in January 1964. The author describes air pollution in Los Angeles and the steps the city has taken to prevent it, including banning home incinerators, forbidding burning in public dumps, and controlling automobile exhaust. Dust, ozone, sulfur dioxide, and nitrogen oxide are discussed separately. Alternatives to fossil fuel-burning cars are mentioned briefly. These include electric propulsion cars and public transportation. The figures are informative and complement the text. Various types of pollution control devices are illustrated.

212 Hagevik, George H. "The Los Angeles Experience," in Woods, Barbara, **Eco-Solutions**. Cambridge, MA: Schenkman, 1972, p. 65-82.

The article discusses the difficulty of enforcing air quality standards and bringing offenders into compliance with particular reference to Los Angeles in the 1960s. The early history of air pollution in Los Angeles is described, beginning with the city's first smog alert in September 1943. The emphasis is on the Los Angeles Air Pollution Control District and its successes and failures in regulating industry and persuading county commissioners to enact tough standards.

213 Haskell, Elizabeth H. **The Politics of Clean Air: EPA Standards for Coal-Burning Power Plants**. New York: Praeger, 1982, 206p.

This is a case study of the U.S. Environmental Protection Agency's procedures to establish pollution control standards for coal-burning power plants built by electric utility companies. A general introduction to air pollution is given. Also discussed are coal use; the utility industry; and the politics of federal regulation. The book proceeds in chronological order from 1971 when clean air standards were set to 1979 when revisions in the standards were finalized.

214 Henderson-Sellers, Brian. **Pollution of Our Atmosphere**. Bristol, England: A. Hilger, 1984, 276p.

The author discusses all aspects of the problem of air pollution, including observation methods, health effects, prevention, legislation, and causes. He emphasizes that pollution cannot be localized. Furthermore, it is difficult to gauge because emissions vary. Changes in technology, fuel usage, mean temperature and rainfall, and legislation all contribute to shifting pollution levels. This is an interesting and enlightening work.

215 Herfindahl, Orris C., and Allen V. Kneese. "Air Pollution," in Herfindahl, Orris C., and Allen V. Kneese, **Quality of the Environment: An Economic Approach to Some Problems in Using Land, Water, and Air**. Washington, DC: Resources for the Future, 1965, p. 24-34.

This is a chapter in a book written by two economists who try to bridge the gap between approaches to environmental problems that are purely economic and that ignore economics entirely. Air pollutants are separated into two categories: primary pollutants such as dust, smoke, and aerosols; and secondary pollutants such as gases produced by the oxidation of hydrocarbons. Secondary pollutants are believed more threatening. Los Angeles is cited as having tremendous secondary pollution. The problems of carbon dioxide and global warming are addressed in a separate section. The comparison of air and water pollution problems is valuable. A short reference list follows the end of the chapter.

216 Kavaler, Lucy. **Dangerous Air**. New York: John Day, 1967, 143p.

The history and possible future of the Earth's atmosphere are discussed in this book. The scenarios presented are not realistic given what is known today about the greenhouse effect, global warming, and ozone depletion. However, the text is of historical interest to individuals curious about how science changes over time.

217 Lafavore, Michael. **Radon: The Invisible Threat**. Emmaus, PA: Rodale Press, 1987, 256p.

This is a guide to what radon is and how to keep your house safe from its effects. Radon is a natural, radioactive gas emitted by uranium in the ground. High levels occur in homes located in areas with large concentrations of uranium in the bedrock. Chapters detail the health effects of radon; testing for radon; eliminating radon; and the position of the government on indoor radon. One useful figure is a map compiled by the Environmental Protection Agency that shows the likely areas of high radon levels in the United States. Included in the book are: a list of places by state to go for help; an inventory of radon testing companies; a glossary; a bibliography; and an index. This is a good source of information for the individual who is concerned about radon in the home.

218 Lave, Lester B., and Gilbert S. Omenn. **Cleaning the Air: Reforming the Clean Air Act: A Staff Paper**. Studies in the Regulation of Economic Activity. Washington, DC: Brookings Institution, 1981, 51p.

The opinion of the authors is that the Clean Air Act has done little to affect pollution. Most of the decrease in air pollution levels is attributable to industries switching fuels for economic reasons. The authors further state that Congress's decision to implement the Clean Air Act through administrative and legal procedures that permit due process resulted in noncompliance. Remedies are suggested, including a complete overhaul of the monitoring and inspection systems.

219 McDermott, Walsh. "Air Pollution and Public Health," in *Scientific American*, **Man and the Ecosphere**. San Francisco: W. H. Freeman, 1971, p. 137-145.

This article first appeared in *Scientific American* in October 1961 and was one of the earliest to try to link air pollution to disease. Unfortunately, a lot of evidence was not yet available in 1961. The author attempts to assess the consequences of long-term exposure to low levels of pollutants, such as carbon monoxide, that are known to be lethal at high concentrations. An interesting aside is that the hazardous level for carbon monoxide is cited as 150 parts per million. Recent studies have shown health problems from as little as 30 parts per million. Thermal inversion, responsible for smog in Los Angeles, is explained. The figures are excellent.

220 Meetham, A. R. **Atmospheric Pollution: Its History, Origins, and Prevention**. New York: Pergamon Press, 1981, 288p.

This is a basic introductory volume on air pollution. Topics include: the origin of fossil fuels; industrial boilers and furnaces; atmospheric pollutants; observation and measurement of pollution; distribution of pollution; effects of pollution; and the prevention of pollution. Legislation also is discussed. The book gives a good overview for the general reader.

221 Miller, E. Willard, and Ruby Miller. **Contemporary World Issues, Environmental Hazards: Air Pollution**. Santa Barbara, CA: ABC-CLIO, 1989, 250p.

This excellent introduction to the problem of global air pollution covers all the major issues, including natural air pollution, photochemical reactions, acid precipitation, carbon dioxide accumulation in the atmosphere, ozone depletion, radioactive contamination, and indoor pollution. A chronology of air pollution events is listed. Also discussed are federal and state laws and legislation. Organizations that combat air pollution are named. A bibliography of books, journal articles, government documents, films, and videocassettes concludes the book.

222 Mueller Associates, SYSCON Corporation, and Brookhaven National Laboratory. **Handbook of Radon in Buildings**. New York: Hemisphere, 1988, 261p.

This is a complete reference designed for both the nonspecialist and the scientist interested in indoor air quality issues. The data are from government and private research studies, conference proceedings, technical journals, and other publications. The book has seven sections that discuss the sources and transport mechanisms of radon, the factors influencing indoor concentrations, health effects, and control methods. A manual for radon in the home is included at the end of the volume. The glossary is extremely useful. This is a good source of specific information on indoor radon.

223 National Research Council, and Assembly of Life Sciences. **Indoor Pollutants**. Washington, DC: National Academy Press, 1981, 537p.

Human beings spend a great deal of time indoors. This thoroughly researched and comprehensive guide to the quality of indoor air, therefore, is indispensable. Subjects include: types of indoor pollution; monitoring; factors that alter indoor pollution; health effects; and control of indoor pollution. This is a lengthy work that should only be used as a reference.

224 Nazaroff, William W., and Anthony V. Nero, Jr. (eds.). **Radon and Its Decay Products in Indoor Air**. New York: John Wiley/Interscience, 1988, 518p.

The book was written by two scientists from the Indoor Environment Program at Lawrence Berkeley Laboratory. The book has six parts, each of which contains a well-organized series of related chapters. Each chapter has an extensive, current reference list. One discussion focuses on the measurement of radon and its hazardous daughter products, showing high indoor radon levels are temporary and can be the result of weather and variations in soil and rock radon concentration. An extensive review of the debate concerning the relationship between lung cancer and radon is also included. Methods for reducing radon levels are reviewed. The last chapter describes the difficulty of identifying communities with elevated indoor radon concentrations. The book is extremely up-to-date and easy to read.

225 Newell, Reginald E., Henry G. Reichle, Jr., and Wolfgang Seiter. "Carbon Monoxide and the Burning Earth," *Scientific American*, v. 261, n. 9, September 1989, p. 82-89.

The authors, preeminent scientists from academia and the National Aeronautics and Space Administration (NASA), designed an instrument to fly on the space shuttle that would measure carbon monoxide levels in the atmosphere. Their findings contradicted earlier assumptions that the gas would be concentrated in the Northern Hemisphere where most of the industrialized nations of the world, and thus, burners of fossil fuels, are located. In contrast, the greatest levels of carbon monoxide are between the equator and 20°S. The implication is that burning of the rainforests and savannahs generates as much of the gas as transportation and industry do. The authors outline the history of detecting carbon monoxide in the atmosphere and describe the instruments used in the investigations. The article is well written and superbly illustrated.

226 Perkins, Henry C. **Air Pollution.** New York: McGraw-Hill, 1974, 407p.

This volume emphasizes specific pollutants: particulates, sulfur dioxide, and nitrogen oxide compounds. Also included are discussions of emissions, the local and global effects of pollution, and the history of air pollution legislation. The author provides a valuable general introduction to combustion, a process with which the reader should be familiar before trying to understand air pollution. The descriptions of meteorology and the dispersion of pollutants from point sources can be technical.

227 Postel, Sandra. **Altering the Earth's Chemistry: Assessing the Risks,** Worldwatch Paper 71. Washington, DC: Worldwatch Institute, July 1986, 66p.

This state of the environment report addresses all aspects of air pollution but emphasizes fossil fuels, acid rain, and atmospheric carbon dioxide. Chapters discuss the risks to food production, the loss of forests, the threats to human health, and methods to minimize risks. Some tables of data are included and the notes are thorough. This is a good overview for the general reader and the concerned individual.

228 Shaw, Robert W. "Air Pollution by Particles," *Scientific American*, v. 257, n. 2, August 1987, p. 96-104.

The application of advanced technological techniques shows that the main source of acidic particles in the atmosphere is fossil fuels. Particles fall in two size ranges: fine, with diameters between 0.1 and 1 micrometer, and coarse, with diameters between 1 and 100 micrometers. The fine particles, which are acidic, derive from combustion reactions. Tests used on the particles include: x-ray fluorescence spectroscopy; x-ray diffraction; and factor analysis. Results indicate that sulfate is the main constituent of the fine-particle cloud in the eastern United States. This is a fascinating article that outlines the observational, experimental, and analysis phases of scientific investigation extremely well.

229 Skjei, Eric, and M. Donald Whorton. "Hazards in the Air," in Skjei, Eric, and M. Donald Whorton, **Of Mice and Molecules.** New York: Dial Press, 1983, p. 91-120.

This is a chapter in a book on technology and human survival. It is listed as a separate entry because the discussion of hazards in the air is a good overview of air pollution. Gases that are described include: carbon monoxide; sulfur oxides; nitrogen oxides; and carbon dioxide. Particulates such as lead, manganese, benzene, and smog also are examined. The sections on indoor pollution and passive smoking are informative and worthwhile. The chapter is well written and scholarly, but it is not technical.

230 Sproull, Wayne T. **Air Pollution and Its Control.** New York: Exposition Press, 1972, 132p.

The technology of air pollution is described in this general book. Topics include: nitrogen and sulfur oxides; carbon monoxide; particulates; and fossil fuels. The legal and economic ramifications of air pollution and its control also are discussed. Although a bit outdated, this is a good introduction to the problem.

231 Teller, Azriel. "Air Pollution Abatement: Economic Rationality and Reality," in Woods, Barbara, **Eco-Solutions**. Cambridge, MA: Schenkman, 1972, p. 49-63.

The essay presents several aspects of the conflict between the desire to have acceptable standards for air quality and the cost to implement them. The author outlines a cost-benefit analysis of economic and social issues. He argues that local government should make the decisions concerning air quality standards because pollution varies from region to region as a result of geography, industry, and the number of automobiles. The different stages of air pollution alerts are described. Diseases that possibly are related to air pollution are mentioned briefly.

232 Tollison, Robert D. **Clearing the Air: Perspectives on Environmental Tobacco Smoke**. Lexington, MA: Lexington Books, 1988, 148p.

This book is a collection of 13 chapters, each written by a different author. Topics are: health effects of tobacco smoke; cigarettes and property rights; smoking in the workplace; smoking and the press; prohibition and third-party costs; and politics of tobacco smoke. The point of the volume was to state the arguments on all sides of the environmental tobacco smoke issue, including the perspectives of health scientists, economists, sociologists, and cigarette industry officials, to allow individuals to make up their own minds. The appendix cites research evidence on environmental tobacco smoke.

233 Treshow, Michael. **Whatever Happened to Fresh Air?** Salt Lake City, UT: University of Utah Press, 1971, 201p.

Written during the first period of environmental concern in the United States, this book provides information on air pollution for the general reader. The author tries to answer several questions: What is air pollution? What are its causes? Where does it originate? Where does it go? What does it effect? What are the legal and financial consequences? How is it legislated and regulated? and How is it controlled? The book suffers somewhat in the modern era from a lack of substantial analysis of ozone and climate change. It is, however, still relevant and timely.

234 Turiel, Isaac. **Indoor Air Quality and Human Health**. Palo Alto, CA: Stanford University Press, 1985, 174p.

In developed countries, more than 80 percent of the air humans breathe is indoor air. The constituents of the air vary markedly from place to place but frequently contain residues from pesticides used to preserve wood, organic sprays and solvents, benzene, formaldehyde from particleboard, and cigarette smoke. The average home's air is worse in the summer because of high vapor pressures and reaction rates. The discussion about energy efficiency in homes and its effect on turnover time for indoor air is illuminating. References include manufacturers of devices designed to alleviate problems and government offices prepared to help.

235 Wagner, Richard. "The Air Around Us," in Wagner, Richard H., **Environment and Man**. New York: W. W. Norton, 1978, p. 427-454.

This is a chapter in a book that gives an excellent introduction to people's impact on the environment. It addresses specific types of air pollution, such as sulfur oxide smog produced primarily by burning coal and photochemical smog made by the interaction of sunlight with fossil fuels burned in automobiles. The effects of each and methods to control the pollution also are outlined. Additional topics covered in less detail include: carbon dioxide and climate; noise pollution; and light pollution. The figures explaining temperature inversions that lead to smog are useful. Suggestions for further reading conclude the chapter.

236 White, Lawrence J. **The Regulation of Air Pollutant Emissions from Motor Vehicles**. Washington, DC: American Enterprise Institute for Public Policy, 1982, 110p.

The author discusses the federal program to regulate emissions from motor vehicles. He argues that a substantial reduction in air pollution was achieved only by imposing large costs on the people who use the vehicles. This is an interesting book about an important part of the federal government's overall attempt to curb air pollution. The emission control program in California is described briefly.

6

Acid Rain

One of the most controversial issues facing industrialized nations is that of acid rain, or more correctly termed, acid deposition, which arises from the emission of sulfur and nitrogen oxides into the air during the combustion of fossil fuels by electrical power plants, factories, and automobiles. The gases mix with water vapor in the atmosphere to produce sulfuric and nitric acids, both corrosive substances, which subsequently settle to the ground hundreds of miles from their point of origin as sleet, fog, rain, or snow. The precipitation is said to be "acid" when its pH drops below 5—the value for natural rainfall. (The pH scale, ranging from 1 to 14, measures the acidity or alkalinity of a solution. A neutral solution has a pH of 7. An alkaline, or basic, solution has a pH between 7 and 14, whereas an acidic solution has a pH between 1 and 7. The farther the number is from 7, the greater the degree of acidity or alkalinity. Ammonia, for example, has a pH of 11. Vinegar has a pH of 3.)

It is an insidious problem: The producers of acid deposition are not the ones who feel its effects. The smokestacks of factories in the northeastern United States, for example, emit the gaseous pollutants high into the atmosphere where the chemicals are dispersed by the winds to settle eventually in the lakes, rivers, and forests of Canada. In Eastern Europe, the waste travels downwind from the industrial areas of the northwest to the agricultural and timber regions of the southeast. Solutions require not only a willingness on the part of those inflicting the damage but also international cooperation. Politics, therefore, play a critical role.

Further hampering positive action on the problem are recent scientific studies that suggest acid deposition in the southeastern United States is not as severe as predicted by earlier investigations. The federal government responded by requiring additional research before the implementation of any severe regulatory measures. This decision was made despite alarming facts from elsewhere in the world: nearly 50 percent of the conifers are dead in the central Alps; the ancient ruins of the Acropolis are suffering damage from acid deposition in Athens; 67 percent of the forests in the United Kingdom and 50 percent of the trees in the former West Germany are dying; 80 percent of the lakes in southern Norway are either endangered or devoid of life; rainfall in parts of China has a pH of 4.0; railroad tracks in Upper Silesia, Poland, are eroding so quickly from acid deposition that speed limits of 25 mph are necessary; and several hundred lakes in Canada have pH values less than 5 (see Walter H. Corson and The Global Tomorrow Coalition's *The Global Ecology Handbook* in chapter 16). Additional effects of acid deposition include: destruction of estuaries and water supplies; corrosion of automobiles; and devastation of agricultural crops.

The sources in this chapter were published in the last decade when the problem of acid deposition was brought to the attention of the general public. Some of the better overviews include: *Acid Rain* by Robert H. Boyle and R. Alexander Boyle; *Acid Rain Information Book, 2d Edition*, edited by David V. Bubenick; "The Human Impact: Acid Rain" by John Gribbon; *Acid Rain: A Plague upon the Waters* by Robert Ostmann, Jr.; and *A Killing Rain* by Thomas Pawlick. John McCormick's *Acid Earth: The Global Threat of Acid Pollution* and the National Research Council's *Acid Deposition: Long-Term Trends* are more technical in their discussions. *The Acid Rain Sourcebook* edited by Thomas C. Elliot and Robert G. Schwieger is an excellent reference for facts and statistics. The controversy surrounding the severity of acid deposition is debated by Julie S. Bach and Lynn Hall in "How Harmful Is Acid Rain?" and by Nigel Dudley and others in *The Acid Rain Controversy*. Additional topics are: the politics and economics of acid rain (note Roy Gould, Chris C. Park, James L. Regens and Robert W. Rycroft, Phil Weller, and Ernest J. Yanarella and Randal H. Ihara); the relationship between technology and acid rain (note James Bellini); the effect of coal plants on acid rain (note James S. Cannon); damage to estuaries (note the Environmental Defense Fund); the acid rain crisis in Canada (note Ross Howard and Michael Perley) and the Appalachians (note Robert Mello); and devastation of trees and crops (note James J. MacKenzie and Mohammed T. El-Ashry, and Sandra Postel). Two bibliographies from the mid-1980s also are cited. New works appear constantly as more people become aware of the crisis and additional research is completed. The perspicacious reader is advised to check the shelves in bookstores and libraries for updated information on this very important issue.

237 Bach, Julie S., and Lynn Hall, (eds.). "How Harmful Is Acid Rain?" in Bach, Julie S., and Lynn Hall (eds.), **The Environmental Crisis, Opposing Viewpoints**. St. Paul, MN: Greenhaven Press, 1986, p. 210-250.

This chapter is in a book that is part of the Opposing Viewpoints series published by Greenhaven Press. The format allows debate of both sides of an issue: two experts argue the pros and cons of one subject. Three topics are included: the acid rain phenomenon is human made; existing technology can control acid rain; and acid rain is a global threat. A lively discussion, this chapter is recommended to the individual interested in all sides of the acid rain problem.

238 Bellini, James. "An Acid World," in Bellini, James, **High Tech Holocaust**. San Francisco: Sierra Club Books, 1986, p. 97-114.

This is a chapter about acid rain in a book that convincingly argues the evils caused by our advanced, technological state. The statistics are alarming: over 50 percent of all car corrosion in Canada is believed the result of acid rain and over $500 million per year in agricultural crops in Europe are destroyed by acid damage. Sulfur dioxide is discussed in detail. The chapter ends on an optimistic note: a program implemented in the 1970s in Japan cut sulfur dioxide emissions and reduced the level of the gas in the air by 75 percent from its all-time high in 1968. The financial cost, however, was considerable. This is a good introduction to a problem with serious repercussions.

239 Boyle, Robert H., and R. Alexander Boyle. **Acid Rain**. New York: Schocken Books, 1983, 146p. ****O**

The serious problem of acid precipitation is addressed in this book. Topics include the causes and extent of the problem, the effects on humans, animals, and our water supply, and the delays of government and industry in prevention. The authors describe action taken in Canada to illustrate the lack of concern in the United States. An index and annotated bibliography augment the text. This is a penetrating look into a problem that is becoming an international emergency.

240 Bubenick, David V. (ed.). **Acid Rain Information Book, 2d Edition**. Park Ridge, NJ: Noyes Publications, 1984, 397p. ****O**

The second edition of the book has been greatly expanded. Topics include the sources of pollutants, methods to lessen the amount of pollution, and regulations enacted by the federal government. An overview of the science of acid rain and the uncertainties in the research is also included. Several ongoing and proposed research studies are described. Although the text can be technical, this is an excellent source of information on any aspect of acid rain.

241 Cannon, James S. **Controlling Acid Rain: A New View of Responsibility**. New York: INFORM, 1987, 53p.

This is a short book that attempts to assign responsibility for acid rain on the users of energy from coal plants in the American midwest. The discussion is valid, but the premise is flawed. The issues in modern society that work together to create (or potentially solve) a problem are interrelated in a more complex way than the author suggests.

242 Curtis, Carolyn (ed.). **Before the Rainbow: What We Know About Acid Rain**. Washington, DC: Edison Electric Institute, 1980, 102p.

This is a collection of information on the acid rain problem compiled by spokespeople for the electric power companies. Included are: the testimony of a member of the Electric Power Research Institute before Congress on acid precipitation; comments on the Environmental Protection Agency's criteria for drafting acid rain legislation; a discussion of the issue by the Edison Electric Institute; and a review of the acid rain problem and its uncertainties by Southern Company Services, Inc. The book is part of the Decisionmakers series that the Edison Electric Institute publishes.

243 Dudley, Nigel, Mark Barrett, and David Baldock. **The Acid Rain Controversy**. London: Earth Resources Research, 1985, 177p.

Most of the scientific work done in Britain prior to the mid-1980s was conducted by the Central Electricity Generating Board, the nation's largest emitter of sulfur dioxide. As a result, many Britons argue that the problem of acid rain and its relationship to air pollution is greatly exaggerated. The nation, therefore, maintains that more research is necessary before any action is taken against acid rain.

244 Elliott, Thomas C., and Robert G. Schwieger (eds.). **The Acid Rain Sourcebook**. New York: McGraw-Hill, 1984, 290p. **F

More than 25 experts contributed to this edited volume of papers that resulted from the First International Conference on Acid Rain held in Washington, D.C., in March 1984. It contains seven sections, each of which is divided into chapters. Topics include the definition of acid rain, current legislation, international problems, emissions reductions, and engineering solutions. An excellent overview, designed to introduce people to the problem, begins the book. The text is accompanied by numerous figures. This book has a wealth of information and is essential reading for the person who wishes to learn the facts.

245 Farmer, Penny (ed.). **Acid Rain and the Environment 1980-1984: A Select Bibliography**. Letchworth, Herts, England: Technical Communications, 1984, 108p.

This is an excellent source of information on acid rain in the early 1980s. The editor has compiled a collection of materials that focus on the United States and American research investigations. The bibliography is not comprehensive but it is a good first step in any research project on acid rain.

246 Fisher, Diane. Prepared for the Environmental Defense Fund. **Polluted Coastal Waters: The Role of Acid Rain**. Washington, DC: Environmental Defense Fund, April 1988, 102p.

The destruction of the nation's estuaries by airborne pollutants is discussed in this research report by the nonprofit environmental group. Although the amount of pollution contributed by acid rain is small in comparison with the tons of chemicals dumped directly into rivers and streams by industrial sources, it does have a noticeable effect. This is an eye-opening account of an important problem.

247 Gould, Roy. **Going Sour**. Stuttgart, Germany: Birkhauser, 1985, 153p.

An overview of the science and politics of acid rain is given in this book. A general introduction details the history of the acid rain problem, the recalcitrance of industry to rectify it, the damage caused by it, and the failure of government to regulate it. The composition, causes, transport, and effects of acid rain are discussed in the second part of the book. The third part of the work concentrates on the politics of the topic. Plates, tables, and figures are collected in an appendix. The short glossary is useful. This complicated problem is simply presented so that the layperson can understand it easily.

248 Gribbon, John. "The Human Impact: Acid Rain," in Gribbon, John (ed.), **The Breathing Planet**. New York: Basil Blackwell, 1986, p. 193-228. **O

This is a collection of eight papers, each written by a different author, that comprises one part in a book on climatic changes. Subjects include: acid drops from fossil fuels; the practice of using warning cones when acid rain is imminent; the volatile mixture of science and politics in the acid rain debate;

the physical and political effects of acid rain; and the cooperation and apathy of the federal government in reducing acid rain. A one-page summary precedes the papers. This is a good, nontechnical overview for the general reader.

249 Howard, Ross, and Michael Perley. **Acid Rain**. New York: McGraw-Hill, 1982, 206p.

The authors discuss the history of the acid rain crisis with emphasis on Canada. In Canada, acid rain has long been recognized as a problem because its effects are more pronounced there than in the United States. The lack of cooperation of American government and industry has thwarted Canadian attempts to remedy the situation. Descriptions of the dead lakes in northern Ontario are shocking. The book is nearly devoid of scientific jargon and is suitable for the general reader. The references and index are thorough.

250 Katzenstein, Alan W. **An Updated Perspective on Acid Rain**. Washington, DC: Edison Electric Institute, 1981, 44p.

This short volume is a general introduction to acid rain. All the important concepts are discussed clearly and concisely in nontechnical language. Topics are: the nature of acid rain; pH; clean air standards; trends of acid rain development; transport of chemical pollutants; consequences of acid deposition; and possible solutions. Prepared for the Edison Electric Institute as a public response document, this is an excellent overview for the general reader.

251 Luoma, Jon R. **Troubled Skies, Troubled Waters**. New York: Viking Press, 1984, 178p.

The author begins by describing a trip through canoe country in the central United States and Canada where lakes hundreds of miles from industrial towns are dead or dying from pollution by acid rain. He also journeys to corporate boardrooms in the two countries to uncover why nothing is done and what the origins of the problem are. In doing so, the author unearths arguments and obstacles placed by industry that may prevent the elimination of acid rain, and conversely, may increase it by the twenty-first century. This is an alarming book that remains objective while being personal and it is suitable for everyone.

252 MacKenzie, James J., and Mohammed T. El-Ashry. **Ill Winds: Airborne Pollution's Toll on Trees and Crops**. Washington, DC: World Resources Institute, September 1988.

This report by the research organization, the World Resources Institute, focuses on the effects of acid rain and ozone on forests and food crops. In the past, the pollutants produced by burning fossil fuels did not travel far from their source. Recently, however, investigations have shown that crop damage and forest destruction is widespread. The rate at which the area effected has grown is alarming.

253 McCormick, John. **Acid Earth: The Global Threat of Acid Pollution**. Washington, DC: International Institute for Environment and Development, 1985, 191p. **S

The first part of this volume is an introduction to the science and logistics of acid pollution and its control. Topics are: the formation of acids; the precipitation of acids; the drainage of acids; and the reduction of acid pollution. The second part emphasizes the global aspect of the problem by discussing the issues affecting the industrialized nations, including the former West Germany, Britain, the former Soviet Union, and Japan.

254 Mello, Robert A. **Last Stand of the Red Spruce**. Washington, DC: Island Press, 1987, 199p.

The author is a Vermont lawyer who grew concerned about the fate of the red spruce forests of the Appalachian Mountains in the eastern United States. He wrote the book during a sabbatical from his job and started with no preconceived notions about acid rain. The book begins by outlining the scientific debate on air pollution and its effect on forests and ends by demanding the reduction of acid rain emissions. The text is alarming and passionate. Sponsored by the Natural Resources Defense Council, the book urges immediate attention to the problem of acid rain to avoid the dire consequences of further environmental damage. The two appendices list significant developments in the history of acid rain. The bibliography is good.

255 National Research Council, Committee on Monitoring and Assessment of Trends in Acid Deposition, Environmental Studies Board, Commission on Physical Sciences, Mathematics, and Resources. **Acid Deposition: Long-Term Trends**. Washington, DC: National Academy Press, 1985, 506p. ****S**

The National Research Council has warned that chemical deposition and atmospheric pollution are the major environmental problems that we face today. Because they are concerned, the NRC has issued a summary of the current state of knowledge about acid rain and chemical input to the atmosphere. Emissions, acid rain, and environmental consequences are discussed. The contribution of industry to the problem is estimated rigorously. This is a scholarly and comprehensive treatment of the acid rain problem that is recommended to the highly motivated and involved reader.

256 Ostmann, Robert, Jr. **Acid Rain: A Plague upon the Waters**. Minneapolis, MN: Dillon Press, 1982, 208p. ****O**

The book examines acid rain and its effects around the world by studying the evidence of the 1970s and by interviewing the researchers who pioneered the investigation of acid precipitation. Possible damage to the environment by acid rain includes the death of thousands of lakes in the northeastern United States, southeastern Canada, Norway, and Sweden, the leaching of toxic metals into the groundwater supply, and the reduction in productivity of the world's forests and crops. The author, a journalist, writes in a style that is casual and informative. Many illustrations accompany the narrative. This is a good source on the acid rain problem.

257 Park, Chris C. **Acid Rain, Rhetoric and Reality**. London: Methuen, 1987, 272p.

Written by a professor at the University of Lancaster in Great Britain, this book covers both the science and politics of acid rain. The discussion is broken into four parts: 1) the problem of acid rain; 2) the science of acid rain; 3) the technology of acid rain; and 4) the politics of acid rain. Each part is similarly separated into subtopics. The explanation of what acid rain is and what it does in the first part of the book is excellent. Tables and figures accompany the text. The bibliography includes both scientific articles and books. This is a well-balanced account suitable for anyone to read.

258 Pawlick, Thomas. **A Killing Rain**. San Francisco: Sierra Club Books, 1984, 206p. ****O**

The focus of this book is on the effects of acid rain on people's lives and the unwillingness of the U.S. government to enact legislation to reduce the pollution. Estimates of the cost of the precipitation hover near $5 billion a year. All moisture on the Earth contains some acid that results from pollution. The urgency of the problem is conveyed effectively. This is a valuable discussion by an investigative reporter in which acid rain is explained clearly and concisely. Appendices list references and the worst polluters in the United States.

259 Postel, Sandra. **Air Pollution, Acid Rain, and the Future of Forests**. Worldwatch Paper 58. Washington, DC: Worldwatch Institute, 1984, 54p.

This is an excellent, nontechnical volume on the effects of acid rain on the world's forests. The author discusses initial symptoms of dying forests, methods to trace pollution back to its source, the economics of acid rain, motor vehicle emissions, the threat to the forest ecosystems, and possible solutions. This alarming and eye-opening account is necessary reading for all individuals concerned about the environment.

260 Regens, James L., and Robert W. Rycroft. **The Acid Rain Controversy**. Pittsburgh, PA: University of Pittsburgh Press, 1988, 228p.

The authors wrote this book to attempt to answer questions they had raised during their discussions about acid rain, such as How did acid rain become a political issue? What is the consensus to its causes and effects? Is a cohesive policy toward acid rain possible? and What is the economic impact? The book is specific and aimed at a high level. Despite the detail, however, it is accessible and informative to the general reader. Figures and tables complement the text. A list of abbreviations, extensive references, and an index are included.

261 Roth, Philip. **The American West's Acid Rain Test**. Washington, DC: World Resources Institute, 1985, 50p.

The author focuses on acid rain in the western United States, a topic that is not discussed frequently. Because weather and soil conditions are different in the West than in the East, acid deposition has a different transport pattern and different environmental impact. Some solutions are suggested. This is an essential source for acid rain in the western United States.

262 Siegel, Mark A., Carol D. Foster, and Nancy R. Jacobs. "Acid Rain," in **Energy: An Issue of the 90's**. Wylie, TX: Information Plus, 1989, p. 113-123.

This is a concise treatment of acid rain that appears as a chapter in a book on energy issues of the 1990s. The discussion begins by answering the questions, "What is acid rain?" and "How is it measured?" Additional topics are: the sources of sulfate and nitrogen in the atmosphere; the effects of acid rain on climate, erosion, trees, and human activity; hazards to aquatic systems;

and damage to soil, vegetation, and human health. Attempts at solutions are summarized. The overview of the international and domestic politics that surround acid rain is provocative. Numerous figures accompany the text. This is a good introduction for the student at any level.

263 Stopp, G. Harry. **Acid Rain: A Bibliography of Research Annotated for Easy Access**. Meutchen, NJ: Scarecrow, 1985, 174p.

This volume contains 884 annotated entries. The annotations are brief. Most of the material included in the bibliography is from the United States or Canada, two of the countries most concerned about acid rain. This is a good source of information with which to begin an investigation of the problem. Many of the works cited, however, are not readily available in most public libraries.

264 Trager, Oliver. "Air and Water Pollution," in Trager, Oliver, **Our Poisoned Planet: Can We Save It?** New York: Facts on File, 1989, p. 2-41.

Acid rain is the emphasis in this chapter on air and water pollution in a book that contains editorials from American and Canadian newspapers. A wide variety of newspapers and viewpoints are represented. Each clipping is in the original typeface of the newspaper in which it appeared and is accompanied by the date it was published. The entries are in a loose chronological order. This unusual format makes for very interesting reading.

265 Weller, Phil. **Acid Rain: The Silent Crisis**. Kitchener, Ontario, Canada: Between the Lines and the Waterloo Public Interest Research Group, 1980, 94p.

In addition to confronting the scientific and environmental aspects of acid rain, the author describes the human consequences. He concentrates on the decisions made by industry and government that exacerbated the acid rain problem. The political and economic ramifications of the decisions and of possible solutions are discussed. This is an interesting and informative book.

266 Yanarella, Ernest J., and Randal H. Ihara (eds.). **The Acid Rain Debate: Scientific, Economic and Political Dimensions**. Boulder, CO: Westview Press, 1985, 345p.

This volume contains numerous papers that have different perspectives on the problem of acid rain. The emphasis is on the policies of various governments, especially the United States and Canada. Western European thoughts also are included. Some papers can be somewhat dry and turgid. This is a good introduction to international opinion.

7

Ozone Depletion, Global Warming, the Greenhouse Effect, and Climate Change

The six warmest years in the history of measurement of global temperatures were in the 1980s. In 1988, the hottest year on record, intolerable heat was accompanied by severe drought during the summer months. Some scientists estimate that global warming of a few degrees Fahrenheit, most of which was absorbed by the oceans, has occurred since the Industrial Revolution of the mid-nineteenth century. Facts like these have convinced many people of the realities and possible consequences of the greenhouse effect and global warming. Temperature increases of only a few degrees will melt the polar ice caps, raising sea level by as much as four feet. Huge areas of the Northern Hemisphere will flood, water supplies in low-lying areas will turn salty, and ocean circulation will change. Warmer temperatures will push climate belts toward the polar regions. Areas already stricken by heat and drought will be devastated.

The temperature of the Earth is regulated in its upper atmosphere primarily by carbon dioxide and secondarily by water vapor, nitrous oxide, and methane. These gases allow sunlight to penetrate to the ground but do not permit the heat reflected from the Earth's surface to escape into space, similar to the panes of glass in a greenhouse. The heat, then, is radiated back to the ground, raising the temperature of the Earth's lower atmosphere by as much as 60 degrees Fahrenheit. This process, known as the "greenhouse effect," allows life to flourish. The danger arises from concentrations of carbon dioxide and other greenhouse gases in the upper atmosphere so large that too much heat is returned to the Earth's surface. Atmospheric concentrations of carbon dioxide were on the order of 275 parts per million (ppm) prior to the Industrial Revolution. That number is now 315 ppm and climbing.

Estimates suggest that nearly 20 billion tons of carbon dioxide are ejected into the air each year. The primary sources of the greenhouse gas are the combustion of fossil fuels by automobiles, coal plants, and oil-fired electrical utilities, the use of chlorofluorocarbons (CFCs), the burning of tropical rainforests, and the application of pesticides in agriculture. The biggest emitters of the gas are the industrialized nations: the United States; the former Soviet Union; Western Europe; and China.

Another environmental issue linked to global warming is the depletion of the Earth's ozone layer, a thin sheet approximately 15 miles above the Earth's surface that shields it from the sun's ultraviolet radiation. The problem came to light in 1982 when a team of British scientists discovered a huge hole in the ozone layer above Antarctica. A similar phenomenon was observed above the Arctic in 1988. Ozone is destroyed by CFCs, the chemicals used in refrigerators, styrofoam packaging, aerosols, air conditioners, and the electronics and computer industries. CFCs react with ultraviolet radiation from the sun in the Earth's upper atmosphere, releasing chlorine in the process. The chlorine is brought by winds to the polar regions where it sticks to the clouds that form in the winter. During the summer, the heat from the sun frees the chlorine, which then destroys the ozone. Loss of the protective ozone allows more of the sun's harmful ultraviolet rays to reach the ground. Ultraviolet radiation damages immune systems, causes skin cancer, and exacerbates eye disease. In addition, CFCs contribute significantly to global warming—one CFC molecule can trap as much as 20,000 times more heat than a carbon dioxide molecule.

Solutions to the problem of global warming and ozone depletion are not easy. Complicated instruments and complex calculations are required to gather and interpret the data that reveal the subtle changes in the Earth's atmosphere. Concentrations of certain gases, ozone and carbon dioxide among them, fluctuate with latitude and season, making determinations of absolute values difficult. Pitfalls abound in the scientific analysis of climate change. Some workers in the field have gone as far as suggesting that the Earth is not getting warmer, but cooler as it enters another Ice Age. These voices, however, grow weaker as more research is done. Even the reasonably conservative Environmental Protection Agency has released the alarming statement that global warming and depletion of the ozone layer is "twice as bad as was thought."

The works in this chapter cover the spectrum of topics associated with ozone depletion, global warming, the greenhouse effect, and climate change from introductory texts on climate modeling (note John Gribbon's *Future Weather and the Greenhouse Effect*) and the carbon cycle (note Bert Bolin) to technical treatments of greenhouse pollutants (note Bert Bolin and others) and polar stratospheric clouds (note Owen Toon and Richard Turco). Numerous excellent overviews exist, including many of the books by John Gribbon and Stephen Schneider, *Fire and Ice* by David Fisher (greenhouse effect and ozone depletion), *Global Alert: The Ozone Pollution Crisis* by Jack Fishman and Robert Kalish (ozone depletion), and *Our Global Greenhouse* by April Koral (greenhouse effect, ozone depletion, and global warming). "Global Climatic Change" by Richard A. Houghton and George M. Woodwell and *The Greenhouse Effect, Climatic Change, and Ecosystems* edited by Bert Bolin and others are somewhat more technical in their approaches. *The Greenhouse Trap* by Francesca Lyman of the World Resources Institute is a good factual reference. Several of the citations discuss possible solutions to the crisis. For opposing points of view, the reader is referred to *Forecasts, Famines, and Freezes* by John Gribbon and *Ice or Fire?* by D. S. Halacy, Jr.

267 Barth, M. C., and J. G. Titus. **Greenhouse Effect and Sea Level Rise**. New York: Van Nostrand Reinhold, 1984, 325p.

This book aims to alert the public and governments of industrialized nations to the economic and environmental effects of a sea-level rise from increased carbon dioxide in our atmosphere. Theoretical scenarios are used to illustrate the main points. The first part of the book assesses the amount by which sea level would rise from predicted global warming. The writing is free of scientific jargon and, as such, the book is an excellent source for the general reader.

268 Bolin, Bert. "The Carbon Cycle," in *Scientific American*, **The Biosphere**. San Francisco: W. H. Freeman, 1970, p. 47-56.

This article was first published in *Scientific American* in September 1970. A firm grasp of the carbon cycle is crucial to understanding the greenhouse effect and its subtleties. The primary carbon cycle from carbon dioxide to living organisms and back again is discussed extensively. The storing of carbon in sedimentary rocks, such as coal, also is described. Additional information includes the daily and annual variations in carbon dioxide in the air, carbon circulation in the oceans, and long-term variations in the carbon dioxide content of the atmosphere. The influence of humans on carbon dioxide in the air is debated at the conclusion of the article. This is an excellent introduction and overview.

269 Bolin, Bert, Bo Doos, Jill Jager, and Richard Warrick (eds.). **The Greenhouse Effect, Climatic Change, and Ecosystems**. Chichester, England: John Wiley, 1986, 541p. ****S**

Although this book was written for researchers in the field, it is also the definitive single-volume guide to the greenhouse effect. The contributors are experts in their fields who prepared the report for the Scientific Committee on Problems of the Environment (SCOPE). All aspects of this important environmental problem are addressed and various perspectives are debated. Any individual with a serious interest in environmental issues will find this an invaluable resource.

270 Breuer, Georg. **Air in Danger: Ecological Perspectives of the Atmosphere**. Cambridge, England: Cambridge University Press, 1978, 189p.

The author, a professor at the University of Vienna, has written a simple and understandable treatment of the threat of pollution in our atmosphere. Five chapters discuss oxygen and carbon dioxide, the history of the atmosphere, the increase in carbon dioxide and its implications for climate change, nitrogen and ozone, and the relationship between the atmosphere and the sea. The consequences of a possible Ice Age also are outlined. The section entitled "Is Our Atmosphere Endangered?" is an excellent introductory essay. The book is translated from the original German by Peter Fabian. A glossary of technical terms, tables, notes, and an index conclude the book. Unfortunately, some of the discussion is outdated.

271 Dorf, Erling. "Climatic Changes of the Past and Present," in Bresler, Jack B. (ed.), **Human Ecology**. Reading, MA: Addison-Wesley, 1966, p. 5-27.

This chapter in a book on human ecology presents an historical review of changes in the Earth's climate. Although the focus is on whether environmental temperatures and other climatic conditions in the past would have allowed humans to flourish, the article is useful as an introduction to changes in climate that were not the result of human tampering with nature. The concepts of climatic change and, in particular, climatic cycles, are critical to the evaluation of theories on global warming. The author states, for example, that today's climate is unusually cold compared with that of 35 to 40 million years ago.

272 Edmonds, Jae, and John M. Reilly. "Energy Consumption and Carbon Dioxide: A Model," in Edmonds, Jae, and John M. Reilly, **Global Energy: Assessing the Future**. Oxford, England: Oxford University Press, 1985, p. 242-264.

This is a chapter in a book that attempts to forecast future demand and supply of energy. Here, the authors describe a model for energy consumption and its effects on carbon dioxide levels in the Earth's atmosphere. The model accounts for differences in production of the greenhouse gas by various fuel types, including nuclear, solar, and hydroelectric power, which contribute little or no carbon dioxide, and coal, oil, and natural gas, which release large amounts of the gas. The model also incorporates long-term effects. The explanation is straightforward; mathematical expressions are used to quantify methods. Numerous flowcharts and graphs illustrate the model inputs and results, respectively. This is an interesting and elegant treatment of a complex problem.

273 Edmonds, Jae, and John M. Reilly, "Global Energy and CO2 to the Year 2050," in Edmonds, Jae, and John M. Reilly, **Global Energy: Assessing the Future**. Oxford, England: Oxford University Press, 1985, p. 265-276.

An overview of global energy use and carbon dioxide production to the year 2050, this is a chapter in a volume that discusses energy supply and demand. The greenhouse effect and global warming are explained briefly. The authors list carbon emissions by fuel type and by geographical source, discovering that the U.S. share of carbon emissions will fall from 25 percent to 20 percent of total carbon dioxide by 2050. This is a consequence of the increased emissions in the developing countries and not a result of reduced consumption of fossil fuels by the United States. Policy responses to the carbon dioxide problem also are discussed. This is a good summary that has several tables and graphs to illustrate the important points.

274 Fisher, David E. **Fire and Ice: The Greenhouse Effect, Ozone Depletion and Nuclear Winter**. New York: Harper & Row, 1990, 232p. ****O**

The book tackles three major environmental problems. The author describes what the greenhouse effect is, how it was discovered, how the clearing of the rainforests contributes to it, and why it is difficult to predict the amount of warming. Also discussed is nuclear winter, which results from the ejection of debris into the sky by the explosions of nuclear warheads. The debris effectively blocks the sun over the entire Earth, causing temperatures to fall dramatically. Although the outlook is bleak, the author maintains his optimism by assuming that actions will be taken before the situation is irreversible. The writing is relaxed. The book is a good overview that everyone will find interesting.

275 Fishman, Jack, and Robert Kalish. **Global Alert: The Ozone Pollution Crisis**. New York: Plenum Press, 1990, 311p. ****O**

The two authors, a NASA atmospheric scientist and a journalist, have written an eloquent account of how ozone pollution is effecting the global carbon cycle. The book is divided into 10 chapters with titles such as "The Concept of Global Change," "The Forests of Oz(one): Ozone's Effects on Forests and Crops," "Governmental Policy: Barking up the Wrong Tree," "South of the Border: The Tropical Ozone Dilemma," and "The 1990s: The Decade We Did Something for Our Atmosphere?" The epilogue describes an optimistic scenario for the year 2004 from the perspective of two graduating college seniors. The text is well written and easy to read. Some graphs are included.

276 Freudenthal, Ralph I., and Susan Loy Freudenthal. "The Greenhouse Effect," in Freudenthal, Ralph I., and Susan Loy Freudenthal, **What You Need to Know to Live with Chemicals**. Green Farms, CT: Hill & Garnett, 1989, p. 146-153.

This is a short essay that is a good introduction and overview of the greenhouse effect. The formation of ozone is explained and the history of the problem is traced. Some discussion of the Environmental Protection Agency's role also is included.

277 Gates, David M. "Carbon Dioxide and Climate Change," in Gates, David M.,**Energy and Ecology**. Sunderland, MA: Sinauer, 1985, p. 144-177.

This is an excellent discussion of the greenhouse effect and the methods scientists use to estimate it. Temperatures from the last 140,000 years on Earth are summarized in an effort to illustrate the pitfalls of this type of analysis. Also described are atmospheric concentrations of carbon dioxide, climate change, climate modeling, fossil fuel futures, and the ecological consequences of increased carbon dioxide. The text is accompanied by several figures and tables. The approach is not technical.

278 Gates, David M. "The Flow of Energy in the Biosphere," in *Scientific American*, **Energy and Power**. San Francisco: W. H. Freeman, 1971, p. 41-52.

An introduction to how incoming solar energy flows through the biosphere (thus maintaining life on Earth) prior to radiating back into space, this article is essential background reading for understanding the greenhouse effect and global warming. Solar energy triggers photosynthesis in plants, which, in turn, remove carbon dioxide from and release oxygen into the air. The addition of carbon dioxide to the atmosphere causes the greenhouse effect. Topics discussed include seasonal variations in solar radiation, relationships between photosynthesis and crop productivity, rates of photosynthesis by agricultural crops and daily energy requirements of people. The figures complement the text. The article first appeared in *Scientific American* in September 1971.

279 Goudie, Andrew. "The Human Impact on Climate and the Atmosphere," in Goudie, Andrew, **The Human Impact on the Natural Environment**. Cambridge, MA: MIT Press, 1986, p. 247-283.

This chapter in a basic text for an introductory course in environmental studies and land resource planning discusses how humans have altered the world's climate by affecting the atmosphere. The author, a professor of geography at Oxford University, writes clearly about subjects as varied as the carbon dioxide problem and the greenhouse effect, aerosols, thermal pollution, vegetation and albedo, population, smoke haze and photochemical smog. Numerous figures are included. One photograph of a doubledecker bus during a smog episode in London in 1952 is particularly shocking. This is a scholarly and useful source of information for the general reader.

280 Green, Fitzhugh. **A Change in the Weather**. New York: W. W. Norton, 1977, 248p.

A former head of the Environmental Protection Agency, the author has written an entertaining account of people's impact, both positive and negative, on the weather. He describes seeding clouds for rain, dispersing fog, and alleviating lightning and thunder as positive actions. Overgrazing of the once fertile Sahara Desert, deforestation of the lands bordering the Mediterranean, and threatening the ozone layer are examples of negative impact. The text is accompanied by some figures. The bibliography and index are extensive. This is an excellent book for individuals interested in weather.

281 Gribbon, John (ed.). **The Breathing Planet**. New York: Basil Blackwell, 1986, 336p.

This collection of papers by noted scientists discusses changes in the climate, both past and present, with an emphasis on the drought-affected areas of the underdeveloped nations. The authors believe that the benefits we achieve by understanding weather and climate are offset by the dangers we create through pollution. The greenhouse effect, logging tropical rainforests, and acid rain are examples of the hazardous practices. Papers are grouped into seven parts: 1) Gaia; 2) the changing climate; 3) cause for concern; 4) the human impact: ozone; 5) the human impact: acid rain; 6) the human impact: the carbon dioxide greenhouse effect; and 7) economics, politics, and climate. Numerous illustrations accompany the text.

282 Gribbon, John. **Forecasts, Famines, and Freezes**. New York: Walker, 1976, 132p. ****O**

The author examines the science of climatology, its attempts to predict short-term weather, and its ability to forecast long-term climates. He also discusses the consequences of some changes in mean global temperatures and rainfalls. For example: a two degree drop in winter temperatures in the United States would increase the yearly fuel bill by $10 billion, whereas no rain in central India would leave 37 million people with no grain to harvest. The concentration is on long-term cooling trends which, in the 1970s, were believed more threatening than possible short-term heating trends created by the burning of fossil fuels. Good illustrations and an index accompany the well-written text. The introduction is an excellent overview for the nontechnical reader.

283 Gribbon, John. **Future Weather and the Greenhouse Effect**. New York: Delacorte Press, 1982, 302p. ****O**

This is an excellent summary of climate and the greenhouse effect written by an astrophysicist from Cambridge University. He discusses both current and past weather patterns and describes the natural cycles of climate that suggest the next ice age is coming. Also explained are the relationships between the sun and climate and between volcanic activity and weather. The last half of the book emphasizes the greenhouse effect and outlines all of the important concepts. Computer models of climate, the basis for most predictions, are described. The author addresses the ramifications of the greenhouse effect for food production and for fossil fuel burning. Are solar and nuclear power the answer? An extensive bibliography, a thorough index, and several figures accompany the text.

284 Gribbon, John. **The Hole in the Sky**. Toronto: Bantam Books, 1988, 192p. ****O**

The author, a distinguished science writer and researcher, explains in nontechnical terms the hole in the Earth's ozone layer. The hole is the size of the United States and is growing. Research is cited that shows the cause of destruction of ozone is CFCs, chemical compounds that are used in a multitude of everyday items, including fast-food packaging, aerosols, and refrigerators. Additional topics are: the effects of increased radiation because of decreased protection from less ozone; experimental studies of the ozone layer; ways to prevent further damage; and greenhouse gases. Suggestions for further reading, an index, and several figures accompany the text. This is a straightforward treatment of an important problem.

285 Gribbon, John. **Hothouse Earth: The Greenhouse Effect and Gaia**. New York: Grove Weidenfeld, 1990, 273p. ****O**

The author, an award-winning science writer, has produced a nontechnical book that discusses global warming and the greenhouse effect in conjunction with the natural climatic cycles of the Earth. He warns that the four degree temperature increase predicted by the year 2030 would melt all the polar ice and turn countries near the Equator into deserts. Specific subjects include: current temperature trends; the natural greenhouse of Earth, Venus, and Mars; the cyclicity of the Ice Ages; carbon cycles; methane production; agriculture; and opposing viewpoints. An appendix summarizes the role of chlorofluorocarbons (CFCs) and ozone in the greenhouse effect. An annotated bibliography and index conclude the volume.

286 Gribbon, John. "The Human Impact: Carbon Dioxide Greenhouse Effect," in Gribbon, John,**The Breathing Planet**. New York: Basil Blackwell, 1986, p. 229-284. ****O**

This collection of eight papers by noted experts is the sixth part in an eight-part book on changes in climate. The topics include: fossil fuel contributions; deforestation; the politics of carbon dioxide; an alternative view to global warming in the next few decades; the relationship among carbon dioxide, climate, and the ocean; current changes in climate that may reflect the greenhouse effect; and natural cycles of climate. Most of the papers contain illustrations and photographs. This is a good introduction to the greenhouse effect that outlines all the scientific arguments.

287 Gribbon, John. "The Human Impact: Ozone," in Gribbon, John, **The Breathing Planet**. New York: Basil Blackwell, 1986, p. 127-192.

One part in a book on changes in climate, this is a collection of eight papers written by respected scientists. Topics include: aerosols and the ozone shield; chlorofluoromethanes and stratospheric ozone; an introduction to the ozone layer; the relationship between ozone loss and cancer; the effect of chlorofluoromethanes on the ozone layer; monitoring halocarbons in the atmosphere; spray cans; and ways to reduce ozone depletion. Some articles are illustrated. Most of the papers are not technical and are suitable for the general reader. Some, however, require a background in basic chemistry.

288 Gribbon, John.**What's Wrong with Our Weather?** New York: Charles Scribner's Sons, 1979, 174p.

The author presents the problem of weather in a lucid and entertaining way. Changes can be expected in the next few decades as weather responds to differences in the activity of the sun. Based on solar research, scientists predict the overall pattern of temperature will cool and the Earth will enter a little ice age similar to that of the early nineteenth century. The ramifications of cooler weather, such as decreased food production, are discussed. Unfortunately, the time of onset of the cooler period is unknown. A full-scale ice age, however, is not predicted until 10,000 years hence. It is important to note that these time scales are much longer than those for the greenhouse effect. This is an informative and scholarly work on climate. The book suffers slightly from not directly confronting global warming and the greenhouse effect.

289 Gribbon, John, and Mick Kelly. **Winds of Change**. London: Headway, 1989, 160p.

A good companion volume to *Hothouse Earth*, which emphasizes the science behind the greenhouse effect, this book describes the scenarios predicted for the twenty-first century by researchers of global warming. The possible situations are not critiqued. Rather, the social and political implications of the changes are discussed. Solutions to potential problems are offered. This is an interesting and alarming account of a probable future world.

290 Halacy, D. S., Jr. **Ice or Fire?** New York: Harper & Row, 1978, 212p.

The author investigates the possibility of human interference to prevent the onset of an ice age in the next century, predicted by climatologists in the 1970s and perceived by citizens as a serious threat after the unusually cold winters of 1977 and 1978. The economics and legal aspects of tampering with the weather are examined. Theories of ice ages, climate, and the effects of cold on humans are discussed. The book is a bit alarmist to an individual with a perspective from the 1990s when emphasis is on global warming and the greenhouse effect. The suggestions for further reading may be outdated.

291 Houghton, Richard A., and George M. Woodwell. "Global Climatic Change," in *Scientific American*, v. 260, n. 4., April 1989, p. 36-47. ****S**

The authors argue that the world is already experiencing a warming trend, citing historical temperature data and the carbon dioxide content of the atmosphere. If current practices of industrialized societies, particularly the burning of fossil fuels and the clearcutting of rainforests, are not strictly controlled, temperatures will increase such that the polar ice caps will melt, sea level will rise, and climate zones will shift. This is a scholarly treatment although not enough is mentioned about long-term global warming and cooling patterns that are part of the natural cycle of the Earth. Excellent figures accompany the text.

292 The Impact Team. **The Weather Conspiracy**. New York: Ballantine Books, 1977, 234p.

This book was compiled from testimonies of experts, scientific studies, and government investigations to inform the layperson about long-term trends in the weather, cited by the Central Intelligence Agency (CIA) in the early 1970s as a major problem facing the United States. The problem stems from variations in food production and human migration patterns that occur when climates change. Topics discussed are: links between civilization and climate; an introduction to climatology; the greenhouse effect; the politics of famine; and energy-saving tips. The debate between global warming and cooling is lucid. The coming of a new ice age, however, receives more attention than the greenhouse effect which shows the remarkable shift in climatology research in the last 15 years.

293 Kellogg, William, and Robert Schware. **Climate Change and Society**. Boulder, CO: Westview Press, 1981, 178p.

This is an excellent overview of climate change and its consequences for modern society. The greenhouse effect is emphasized. The text is straightforward and free of jargon, but, unfortunately, is slightly outdated. An extensive reference list is included.

294 Koral, April. **Our Global Greenhouse**. New York: Franklin Watts, 1989, 64p. ****O**

The book is geared toward high school science students but is also a good introduction for the general reader. The primary emphasis of the book is on sources of excess carbon dioxide in the Earth's atmosphere, such as the burning of fossil fuels and the release of chlorofluorocarbons, and the resultant greenhouse effect. Other topics include: temperature changes; ozone depletion; and sea-level rise. The illustrations are colorful, attractive, and informative. Simple experiments for the individual to do are outlined.

295 Landsberg, H. E. "Trends in Climatology," in Bresler, Jack B. (ed.), **Human Ecology**. Reading, MA: Addison-Wesley, 1966, p. 28-48.

A history of the study of climate is summarized in this article. Important concepts also are described. Topics include: the energetics of climate; the classification of climates; change in climate; and microclimatology. Carbon dioxide and its relationship to the greenhouse effect and global warming are discussed in the context of cycles that last thousands of years. One interesting note is the effect of cities on climate. Cities, for example, are hotter and cloudier than rural areas. The article is not quantitative and is suitable for anyone who wants background information on climate.

295a Lyman, Francesca. **The Greenhouse Trap**. Boston: Beacon Press, 1990, 190p. **F

Do you want to know what to do about global warming and greenhouse gases? If so, suggestions are at your fingertips in this volume put together with the cooperation of the World Resources Institute, an independent center for policy research on natural resources and the environment. Facts abound in discussions of the evolution of the Earth's climate, the effect of fossil fuels and technological practices on the atmosphere, the science of climatology, and the actions of governments. The writing is lucid and informative. Tips on what the individual can do are shaded in gray and accented with bold type. A short bibliography is included. This excellent overview is recommended enthusiastically. Notes on the inside back cover indicate two additional volumes on deforestation and energy are in preparation. The reader should keep an eye out for them.

296 McKibben, Bill. **The End of Nature**. New York: Random House, 1989, 226p.

The author has written a straightforward, impassioned account of the environmental threats that are facing the planet. His emphasis is on the greenhouse effect and its relationship to global warming and climate change. The argument is made that global warming, the destruction of the rainforests, and pollution in general are the consequences of people's intellectual belief that they are not a part of the natural world. This tenet must be discarded if the planet is to be saved.

297 Miller, Alan, and Irving Mintzer. **The Sky Is the Limit: Strategies for Protecting the Ozone Layer**. Washington, DC: World Resources Institute, 1987, 43p.

The authors review the benefits and risks of ozone modification. The techniques for eliminating, or at least reducing, chlorofluorocarbons (CFCs) are described. The authors argue that alternatives to hazardous substances are available and are inexpensive. The cost to protect the ozone layer against further depletion is modest. As with everything else in our modern society, however, the costs will rise the more we delay. The authors, therefore, urge politicians and the public to act quickly.

298 Mintzer, Irving M. **A Matter of Degrees: The Potential for Controlling the Greenhouse Effect**. Washington, DC: World Resources Institute, 1987, 60p.

This is an excellent and lucid overview of the greenhouse effect and the various models used to predict and analyze it. The report contains six parts: 1) an introduction to the report; 2) the greenhouse problem; 3) the model of warming commitment; 4) scenarios of future energy use and carbon dioxide emissions; 5) atmospheric build up of carbon dioxide and other greenhouse gases; and 6) modeling results and policy implications. Fourteen tables are collected in an appendix. Important points are highlighted with illustrations that help to clarify technical issues. Good notes accompany the text.

299 Moomaw, William R., and Irving M. Mintzer. **Strategies for Limiting Global Climate Change**. Washington, DC: World Resources Institute, 1989, 30p.

The senior author, the director of the Climate, Energy and Pollution program at the World Resources Institute, and his co-author have prepared an agenda to slow the increase of greenhouse gases in the Earth's atmosphere. The first of their strategies is better fuel efficiency, essential in an automobile-driven society like the United States.

300 Oort, Abraham. "The Energy Cycle of the Earth," in *Scientific American*, **The Biosphere**. San Francisco: W. H. Freeman, 1970, p. 13-23.

This article first appeared in *Scientific American* in September 1970. The emphasis is on solar energy, which is absorbed by the Earth and then radiated into space as heat. The circulation of the atmosphere and the oceans distributes solar energy over the surface of the Earth. Seasonal and daily variations in solar energy and the adjustments made by the atmosphere are discussed thoroughly. The article concludes with the possible effects human intervention could have on the circulation pattern. The text is well written and superbly illustrated.

301 Oppenheimer, Michael, and Robert H. Boyle. **Dead Heat: The Race Against the Greenhouse Effect**. New York: Basic Books, 1990, 268p.

The book, written by a senior scientist for the Environmental Defense Fund and a science writer, traces the history of human addiction to fossil fuels and outlines the consequences for the future should this reliance on hydrocarbons continue. The authors urge Americans to change their ways before it is too late. Assessing the time table in which action must be undertaken also is part of the discussion. Issues include: development in the Third World; alternative energy sources; and removal of the military-industrial economy. Notes and an index conclude the book.

302 Pearman, Graeme (ed.). **Greenhouse: Planning for Climate Change**. Leiden, Netherlands: E. J. Brill/CSIRO, 1989, 752p.

This is a collection of scientific papers presented at a conference on the greenhouse effect in Melbourne, Australia, in 1987. Although some of the papers are quite technical and are aimed at the specialist, the volume is included in the bibliography because it emphasizes the Southern Hemisphere.

303 Plass, Gilbert N. "Carbon Dioxide and Climate," in *Scientific American*, **Man and the Ecosphere**. San Francisco: W. H. Freeman, 1971, p. 172-179.

This article first appeared in *Scientific American* in July 1959 and is relevant today in discussions on the greenhouse effect and global warming. The author points out that although carbon dioxide is not a pollutant that presents a direct health threat, it is a pollutant that can alter natural cycles. He argues that the global carbon cycle already has been perturbed and, consequently, so has the world's climate. The study of carbon dioxide and other "greenhouse gases" is ongoing; many of the numbers quoted in this article have been revised several times since 1959. In addition, some scientists do not believe that the carbon cycle has been affected at all. Nonetheless, this article is an excellent source for an introduction to the problem. The figures are excellent.

304 Revelle, Roger. "Carbon Dioxide and World Climate," in *Scientific American*, v. 247, n. 2, August 1982, p. 35-43.

Written by a Professor at the University of California at San Diego, and former president of the American Association for the Advancement of Science, this article is lucid and straightforward. The author discusses three questions: 1) How much carbon dioxide will be added to the atmosphere in the years ahead?; 2) Will the increase result in a change in climate and a rise in temperature?; and 3) Will these changes have major effects on modern societies? The history of the "carbon dioxide question" is traced. All important concepts are explained simply and illustrated where appropriate.

305 Roan, Sharon L. **Ozone Crisis**. Wiley Science Editions. New York: John Wiley, 1989, 270p.

The author, a staff science writer for the *Orange County Register* in southern California, has written a clear account of the history of the ozone crisis by focusing on the experiences of F. Sherwood Rowland and Mario Molina, the two scientists who tried to warn that chlorofluorocarbons (CFCs) were destroying the globe's ozone layer. Despite a ban on CFCs in the late 1970s, the ozone crisis largely was ignored until the discovery of a hole in the ozone layer above Antarctica in 1985. The author concludes with summaries of actions taken by world governments to avert further erosion of the ozone layer. The story is presented in chronological order with each of the 16 chapters corresponding to a period of months between 1973 and 1988. A short bibliography, index, and timeline are included.

306 Schneider, Stephen H. "The Changing Climate," in *Scientific American*, v. 261, n. 9, September 1989, p. 70-79.

The author argues that global warming will be significant enough to detect by the early twenty-first century and stresses that cuts in emissions will slow the onset of higher temperatures by reducing the accumulation in the atmosphere of heat-trapping gases such as carbon dioxide. The greenhouse effect is described clearly. Fluctuations in climate through geologic history also are discussed. The article benefits from a lucid presentation of the relationship between atmospheric carbon dioxide and temperature. Although the author asserts global warming is a serious threat to our planet, others disagree. This point is not addressed adequately. The article is well written and very informative.

307 Schneider, Stephen H. "Climate Modelling," in *Scientific American*, v. 256, n. 5, May 1987, p. 72-89.

The author, an atmospheric scientist at the National Center for Atmospheric Research, has written an excellent introduction to the science and mathematics of climate modelling. All the basic concepts and subtle nuances are explained. The Cretaceous Period (a time about 100 million years ago when the Earth was tropical), the modern greenhouse, and nuclear winter are discussed. The illustrations are superb.

308 Schneider, Stephen H. **Global Warming: Are We Entering the Greenhouse Century?** San Francisco: Sierra Club Books, 1989, 317p. ****O**

A scientist at the National Center for Atmospheric Research, the author has written a nontechnical book that explains the controversies around global warming models and discusses the implications of public policy. His analysis of possible errors within the computer models is particularly illuminating. The conclusion is that we are entering the greenhouse century, although the author does acknowledge that some unknown mechanism may reduce or prevent global warming. This is an informative and scholarly text that everyone will enjoy.

309 Schneider, Stephen H., and Randi Londer. **The Coevolution of Climate and Life**. San Francisco: Sierra Club Books, 1984, 563p.

This is a fact-filled volume about the relationship between climate and the evolution of life written by a scientist at the National Center for Atmospheric Research. The approach is even-handed, particularly when discussing theories that may not have universal support. The style is nontechnical. This is an excellent resource for the general reader.

310 Schneider, Stephen H., with Lynne E. Mesirow. **The Genesis Strategy: Climate and Global Survival**. New York: Plenum Press, 1976, 419p.

This is an accessible introduction to climate and possible human effect on it written by a noted scientist and scholar. The complex interactions among climate, food, population, technology, energy, pollution, and politics are explored. The Genesis Strategy of the title refers to the biblical story of Joseph who stored harvests from seven plentiful years to last through years of famine. The authors stress that human activity is altering climate and that grain reserves are too low to endure lean years. In a more general sense, they argue that diversity and margins of safety should apply to all human endeavors, especially those that affect our environment and our future. This is an interesting, authoritative, and well-written book. It is still timely despite having been written in the mid-1970s. Extensive notes, references, and index are included.

311 Thomson, William A. R. **A Change of Air: Climate and Health**. New York: Charles Scribner's Sons, 1979, 225p.

A discussion of the effects of climate on health, this book illustrates the dynamics of changing climates and their consequences. The author describes heat, cold, wind, high altitude, sunshine, and sea breezes. He also examines the relief of illness such as rheumatism, migraines, mental health, cancer, and lung disease by certain climates. The chapter on human-made climates is particularly worthwhile. The bibliography is short but provides the interested reader with several additional sources.

312 Toon, Owen B., and Richard P. Turco. "Polar Stratospheric Clouds and Ozone Depletion," in *Scientific American*, v. 264, n. 6, June 1991, p. 68-75. **S

The authors, experts in the field of atmospheric science, describe the formation of strange, almost invisible clouds in Antarctica during the winter. These clouds, composed of ice and frozen nitric acid, conspire with chlorofluorocarbons (CFCs) to produce the ozone hole noted every spring in the atmosphere above the Arctic and Antarctic. The chemistry of the clouds prevents the removal of free chlorine, an ozone-destroying element given off during the breakdown of CFCs. The article is beautifully illustrated. This is an interesting and informative treatment of an important problem that occurs naturally but is exacerbated by humans.

313 U.S. Congress, Office of Technology Assessment. **Catching Our Breath: Next Steps for Reducing Urban Ozone**. OTA-O-412, Washington, DC: U.S. Government Printing Office, July 1989, 237p.

As preparation for new legislation on the Clean Air Act, the U.S. Congress requested this report on urban ozone, the most difficult pollutant to control. Subjects include: the health effects of ozone; the effects of ozone on crops and forests; lowering ozone; controlling emissions of volatile organic compounds; and policy options. The first chapter is a summary of what ozone is and why it is necessary to control. Numerous figures, graphs, and tables accompany the simple text. This is a good source of factual information about ozone levels in our cities.

314 U.S. Environmental Protection Agency. **The Potential Effects of Global Climate Change on the United States**. Washington, DC: U.S. Government Printing Office, October 1988, 3 volumes, n.p.

An excellent overview of the potential problems that global warming may cause in the United States, this is a draft report prepared for Congress that may be difficult to find in some local libraries. The Environmental Protection Agency discussed the loss of coastal wetlands due to sea-level rise, the increase in electricity use in the South, deforestation, and the rise in extinction rates for endangered species that would arise with a 5 to 8 degree change in temperature.

315 Weiner, Jonathon. **The Next One Hundred Years**. New York: Bantam Books, 1990, 312p.

The author describes the effects of global warming over the next century. Deforestation of tropical regions, ozone depletion, and the greenhouse effect are discussed. The problems and pitfalls of scientific modeling are explained in a nontechnical way to illustrate the difficulty in determining the amount of depletion of ozone or warming of the Earth. The author maintains, however, that despite this, the scientific community is in agreement over global warming—and this is not absolutely accurate. The book is eloquently written and captivating. The author manages to remain optimistic despite the critical condition of the environment.

316 Young, Louise B. **Sowing the Wind: Reflections on the Earth's Atmosphere**. New York: Prentice-Hall, 1990, 207p.

The author, a physicist with a master's degree in geophysical sciences from the University of Chicago, has written a lyrical account of the Earth's atmosphere, exposing its necessity, its beauty, and, above all, its fragility. The book is divided into two parts: "This Excellent Canopy" and "Assessing the Damage." The first describes the parts of the atmosphere, weather and ocean circulation patterns, the influence of the atmosphere on the biosphere, and solar energy. The second part outlines environmental problems, including global warming and the greenhouse effect, acid rain, and ozone depletion. An index, notes, and suggestions for further reading conclude the book. The text is free of scientific jargon and is a joy to read.

8

Water Pollution

Two-thirds of the Earth's surface is covered with water. At first glance, water is a resource that seems almost infinite as the blue oceans extend into the horizon, rivers meander through the woods, and lakes populate the countryside. In reality, pollution threatens the majority of the world's water and marine ecosystems. Believing that "dilution is the solution," countries have dumped billions of tons of sewage, industrial waste, radioactive material, and medical debris into the oceans over the years. Pollution, however, is not restricted to the world's oceans. Lakes, streams, rivers, groundwater, and coastal zones including salt marshes, tidal wetlands, estuaries, and coral reefs are affected. Acid precipitation adds corrosive chemicals to lakes and agricultural runoff contributes the toxic ingredients in pesticides to streams and groundwater supplies. One estimate suggests that if all the Earth's water were put into a gallon container and the undrinkable part removed because it was too salty, polluted, or frozen, only one drop would remain (see *Design for a Livable Planet* by Jon Naar in chapter 16). Overpopulation, especially in arid regions, stretches that one drop.

The sources of pollution differ in the various aqueous environments. The primary culprits in devastation of the coastal zones are municipal sewage, industrial waste, urban runoff, and airborne pollutants such as the emissions from automobiles, factories, and power plants. In the open ocean, the most serious threats come from the dumping of sewage sludge, industrial waste, and material dredged from harbors and rivers to keep them clear for ships, the spilling of oil from tankers and offshore oil platforms, and the discarding of garbage and plastic from boats that routinely traffic the seas. Another problem for the marine environment is the overharvesting of fish and the overexploitation of whales.

Major sources of pollution to freshwater supplies include industrial and mining operations, hazardous waste disposal, agricultural practices, and acid rain. Contaminants released into the environment by these processes run the gamut from toxic metals like lead, cadmium, and mercury to organic compounds such as petrochemicals, phenols, and dioxin. What is disconcerting is that a study published in *The World Rivers Review* in 1989 reported that 250 city sewage facilities and over 600 industrial sites regularly dumped toxic waste into the freshwater system of the United States.

The most important and vulnerable water supply of all, groundwater, is the least regulated and may be the most contaminated. Estimates indicate that groundwater is the only source of drinking water for the majority of the people in the United States and that as much as a quarter of the supply is unusable. Groundwater pollution arises from the infiltration of salt water into aquifers, the proliferation of disease-causing bacteria in contaminated water, and the disposal of toxic and radioactive waste beneath the surface. Because groundwater can flow for great distances, hazardous material that is buried beneath the ground in one spot can contaminate water wells several miles away.

The topic of water pollution is immense and a tremendous number of works on various aspects of the problem exist. This chapter does not attempt to summarize them all but does include select publications that discuss all the important issues. Texts range from the scholarly treatment of groundwater contamination by Linda Abriola (*Groundwater Contamination*) to the more general introduction to threats to the freshwater supply by William Ashworth (*Nor Any Drop to Drink*). Good overviews include: "The Last Refuge" by James Bellini, the dated yet still timely *Death of the Sweet Waters* by Donald E. Carr, "Our Most Precious Resource: Water" by Thomas Y. Canby, *Water Pollution* by George Laycock, *The Oceans* by Wesley Marx, *Neptune's Revenge: The Ocean of Tomorrow* by Anne W. Simon, "Hazards in the Water" by Eric Skjei and M. Donald Whorton, "Air and Water Pollution" by Oliver Trager, and "Water Pollution" by Richard Wagner. Other subjects are lead in the marine environment (note Marko Branica and Zdenka Konrad), the Great Lakes (note William Ashworth, and Charles F. Powers and Andrew Robertson), water shortages in the western United States (note Charles Bowden and Marc Reisner), the manipulation of water resources (note Robert Boyle and others), oil in the marine evironment (note Joseph E. Brown and William H. MacLeish), dolphins (note Kenneth Brower), thermal pollution (note John R. Clark), pollution monitoring (note Edward D. Goldberg), hydrology (note publications by the International Association of Hydrological Sciences), cleanup and management (note E. W. Kenworthy and Allen V. Kruse), fertilizers and pesticides (note the Organisation for Economic Co-operation and Development), waste disposal and groundwater (note R. Perry), the politics of the ocean (note Edward Wenk, Jr.), and experiments the individual can do at home to determine the extent of pollution in water (note D. I. Williams and D. Anglesea).

317 Abriola, Linda M. (ed.). **Groundwater Contamination**. Washington, DC: International Association of Hydrologists, Special Publication no. 185, 197p. **S

This volume contains the proceedings of a symposium held in Baltimore, Maryland, in May 1989 as part of the American Geophysical Union's annual spring meeting. The papers are divided among three categories: 1) processes governing the transport and fate of contaminants; 2) aquifer degradation from agricultural practices; and 3) rehabilitation projects. Most papers emphasize mathematical modeling. A strong background in college-level mathematics is recommended. Papers in the last third of the book are less quantitative and are more suitable for the general reader.

318 Ambroggi, Robert P. "Water," in *Scientific American*, v. 243, n. 3, March 1980, p. 100-117.

This well-illustrated, scholarly, and lucid article is in a special issue of *Scientific American* on economic development written by a senior advisor to the United Nations Food and Agriculture organization. The author states that ensuring an adequate supply of water for agriculture, industry, and people requires human intervention. He discusses global reserves of fresh water, the hydrologic cycle, major river basins, the uses of water, and irrigation.

319 Ashworth, William. **The Late, Great Lakes**. New York: Alfred A. Knopf, 1986, 274p.

The author chronicles the history of the Great Lakes from the Ice Age through the time of European settlement to the modern era. The "fifth coast"—the Great Lakes provide the United States with 5,000 miles of continuous coastline and hold one-fifth of the world's freshwater supply—as he calls it, is awash with toxic chemicals and barren of most aquatic life. PCBs, chromium, lead, oil, and DDT are common in harbor sediment. Love Canal, which drains into the lakes, and the Cuyahoga River, once so polluted it caught fire, are discussed. The author has written a powerful and scathing commentary on the apathy and exploitive behavior of the people managing the lakes. The bibliography, index, and maps on the inside covers are useful.

320 Ashworth, William. **Nor Any Drop to Drink**. New York: Summit Books, 1982, 272p. ****O**

The author estimates that within the next few years, all but eight states will have water shortages. He cites the influx of people to the desert southwest, the increased output of hazardous waste by industry, and the drop in water tables as the major factors. The book contains many horror stories of water shortages in major urban areas throughout the 1970s and 1980s. Although the water shortage may seem overemphasized at times, it is a serious problem in metropolitan regions. The third part of the book describes pollution of water supplies by radioactive elements, carcinogens, salt, and acid rain. This is a thought-provoking book that may stir one to action. It has a good bibliography (slightly outdated) and index.

321 Barrett, Spencer C. H. "Waterweed Invasions," in *Scientific American*, v. 262, n. 4, October 1989, p. 90-97.

Two aquatic weeds, the water hyacinth and the kariba weed, are described to illustrate the changes that occur in ecosystems disturbed by human activity. The two plants can grow and multiply rapidly, thereby choking a thriving water community in a single growing season. The plants feed on nutrients supplied by agricultural activity. Pesticides used to kill the plants also destroy other aquatic life, making the use of the chemicals counterproductive. The author discusses the reproductive cycle of the two weeds in detail. Figures are excellent. A map of the distribution of the plants and the dates when the water hyacinth was introduced to the region is particularly informative.

322 Bellini, James. "The Last Refuge," in Bellini, James, **High Tech Holocaust**. San Francisco: Sierra Club Books, 1986, p. 130-164. ****O**

This chapter in a book that emphasizes the problems caused by modern technology is a good introduction and overview of global water pollution of all kinds. The chapter begins with a description of the collision between a cargo ship and a car ferry off the Belgian coast in 1984. Aboard the ship were 450 tons of uranium hexafluoride gas and 500 pounds of plutonium, some of which spilled into the North Sea. A table that lists the type and origin of contamination by industry is particularly informative. This alarming essay depicts the decline in the world's water quality as critical.

323 Bowden, Charles. **Killing the Hidden Waters**. Austin, TX: University of Texas Press, 1977, 174p.

The essay concerns itself with the consequences of groundwater exploitation in the desert southwest of the United States and northwest of mainland Mexico, an area known historically as Pimeria Alta. The author traces the development of the Sonoran Desert from the late nineteenth century to the present and illustrates the interaction between people and water in the region. The Texas plains also are discussed. The book is divided into three parts. The first two discuss geographic areas. The third, titled "Taking Care of Business," concentrates on the debate over water supply and human demand. The essay is well written and easy to read. The bibliography is an excellent reference for the history and environment of the desert southwest of the United States.

324 Boyle, Robert H., John Graves, and T. H. Watkins. **The Water Hustlers**. San Francisco: Sierra Club Books, 1971, 253p.

The authors examine the manipulation of water resources in Texas, California, and New York. Plans in Texas include the impounding of water in every stream and the siphoning of additional supply from the Mississippi. In California, the discussion focuses on the battle between the thirsty south and the wet north. The leaders in New York wanted to build numerous dams near the headwaters of the Hudson River. This is an interesting book that elucidates the short-sightedness of many involved in urban development.

325 Branica, Marko, and Zdenka Konrad (eds.). **Lead in the Marine Environment**. Oxford, England: Pergamon Press, 1980, 353p.

This volume records the proceedings of the International Experts Discussion on Lead Occurrence, Fate and Pollution in the Marine Environment held in Yugoslavia in 1977. It discusses the chemistry of lead and the ways in which it is released into the environment. Topics include: types of lead compounds and methods to detect lead; distribution of lead and the relationship to human inputs; lead metabolism; effects of lead on the cells of mammals; and toxicity. The collection of papers is technical and aimed at the scientific community. The volume, however, gives a good overall picture of lead pollution in the marine environment as the problem is perceived by an interdisciplinary group of scientists.

326 Brower, Kenneth. "The Destruction of the Dolphin," in *The Atlantic Monthly*, v. 263, July 1989, p. 35-58.

The article describes the modern tuna industry and its practice of entangling (thereby killing) dolphins in its nets. The author tells the story of Sam La Budde, an environmentalist at the Earth Island Institute, who worked aboard a tuna fishing boat gathering data about dolphin killings. Sam brought an 8mm camera on board ship, where he worked as the cook, allowing him time to film the tuna harvest. The descriptions of the catch are harrowing and graphic. This is a well-written and informative piece with a distinct environmental bias.

327 Brown, Joseph E. **Oil Spills: Danger in the Sea**. New York: Dodd, Mead, 1978, 123p.

The author cites some alarming facts: 6.113 million metric tons of oil per year enter the marine environment and 13,000 spills per year occur in the territorial waters of the United States alone. This book discusses the devastating effects of oil spills, cleanup techniques, research into the fate of oil in the sea, and alternative energy sources. The text is simple and accompanied by numerous black-and-white photographs. Because it was written in the late 1970s, the book is a little outdated.

328 Bulloch, David K. **The Wasted Ocean**. New York: Lyons & Burford, 1989, 150p.

The author, a past president of the American Littoral Society (a nonprofit organization dedicated to preserving the coastlines of the United States), has written an alarming account of the polluted marine environment for the general reader. He explains how and why we abuse our oceans. Most importantly, however, the author includes a pragmatic guide, summarized in chapter 7, "The Aquaphile's Agenda," to stopping the pollution. It is the belief that the dire situation can be changed that makes this frightening book optimistic. The appendices describe important federal statutes and regulations, list federal coastal management bodies, and name environmental organizations working toward saving the coastline.

329 Canby, Thomas Y. "Our Most Precious Resource: Water," in *National Geographic*, v. 158, August 1980, p. 144-179. ****O**

The state of the nation's water supply is assessed. Stories are drawn from the wet East where only 12 percent of the water used each day is not returned and from the dry West where the figure exceeds 50 percent. Eutrification of Lake Tahoe, the burning of the Cuyahoga River in Cleveland, acid rain in the northeast and land subsidence in Texas from water extraction are among the topics discussed. This is an excellent overview of the problems of pollution and the scarcity of good water.

330 Carr, Donald E. **Death of the Sweet Waters**. New York: W. W. Norton, 1966, 257p. ****O**

The author warned about the destruction of the marine environment decades before the current crisis. All aspects of water pollution known in the early 1960s are discussed. A notable omission is the effect of mercury effluents. An alarming and poignant account of a problem that persists and grows worse, this book, although outdated, is still relevant today.

331 Clark, John R. "Thermal Pollution and Aquatic Life," in *Scientific American*, **Man and the Ecosphere**. San Francisco: W. H. Freeman, 1971, p. 162-171.

This article, which first appeared in *Scientific American* in March 1969, discusses temperature in water and the effects on marine life of its rise or decline. Temperature is viewed by ecologists as the primary control of life. One consequence of higher temperatures is the increased demand on dissolved oxygen. Changes in temperature, therefore, can be lethal. Sources of thermal pollution in bodies of water include electric and nuclear power plants. Possible uses for the waste heat generated by the power plants are presented. Two factors suggest use of the heat is unrealistic: 1) the temperature is too low for the heat to be economically profitable; and 2) power plants would have to be in urban areas if the heat is to be used.

332 Coffel, Steve. **But Not a Drop to Drink: The Lifesaving Guide to Good Water**. New York: Rawson, 1989, 323p.

The author, a staunch environmentalist, paints an alarming picture of the future of our nation's water supply. The most widespread pollutants are described as are the worst and best water supplies in the country. On a more practical note, suggestions are given to improve your home drinking water. Home filtering and distillation systems are critiqued. Lists of contaminated sites by states, superfund sites, water information sources, and state water quality and health agencies are in the appendices.

333 The Conservation Foundation. "Natural Resources: Land, Water, and Energy—Water," in The Conservation Foundation, **State of the Government: A View Toward the Nineties**. Washington, DC: The Conservation Foundation, 1987, p. 223-234.

This discussion of water as a natural resource is in a chapter of a report by the Conservation Foundation designed to allow informed decisions to be made about existing and potential environmental issues and problems. Trends in water use and sources of water are described. Several figures accompany the clear and concise essay. This is an excellent summary recommended to everyone.

334 The Conservation Foundation. "Traditional Environmental Contaminants—Water Quality," in The Conservation Foundation, **State of the Government: A View Toward the Nineties**. Washington, DC: The Conservation Foundation, 1987, p. 87-105.

This section of a chapter in the Conservation Foundation's report on the state of the environment is an excellent overview of surface water and groundwater quality. Types of pollutants described include: dissolved oxygen; bacteria; suspended solids; metals and toxic substances; nitrates; and radionuclides. The sources of water quality problems also are discussed. The text is accompanied by numerous figures, graphs, and tables. The overview is recommended to the individual concerned about water pollution but who wants to review the statistics behind the problem.

334a Davidson, Art. **In the Wake of the Exxon Valdez: The Devastating Impact of the Alaska Oil Spill**. San Francisco: Sierra Club Books, 1990, 332p.

The author, a natural resource planner for the state of Alaska, began this book the day he heard the news of the 10 million-gallon oil spill from the *Exxon Valdez* in Prince William Sound. He interviewed locals who helped with cleanup efforts, volunteers who traveled from all over the world, Native Alaskans, federal and state government officials, environmental scientists, and Exxon executives. The result is a personal story told in three parts: 1) the spill, a detailed minute-by-minute account of the disaster; 2) the response, a description of the damaged ecosystems, destroyed wildlife, and attempted rescue; and 3) the wake, an outline of the immediate and lasting effects on the Alaska wilderness, the oil companies, and the native populations. Good notes and an index accompany the text.

335 Fair, Gordon Maskew, John Charles Geyer, and Daniel Alexander Okun. **Water and Wastewater Engineering, Volume 2: Water Purification and Wastewater Treatment and Disposal**. New York: John Wiley, 1968, 480p.

This text, containing 20 chapters, considers physical, chemical, and biological processes in the management of water. The book was written for engineers and is very technical. The first four chapters discuss water quality, purification, and wastewater treatment. The next five chapters focus on physical processes in and properties of water. The five following chapters emphasize biological aspects of water treatment. The last two chapters address engineering decisions related to the design of treatment plants. The book assumes familiarity with college mathematics and basic engineering. It is useful to people who care about the nuts and bolts of water treatment.

336 Friedmann, Wolfgang. **The Future of the Oceans**. New York: George Braziller, 1971, 132p.

This book is written by a professor of international law who presents the view that the exploitation of the oceans without international compromise and cooperation will end in military, political, and economic confrontation and that the competition for resources will produce further pollution of the marine environment. The last chapter in the book reproduces the United Nations resolutions of December 17, 1970, on the principles governing the seabed and ocean floor, and the subsoil thereof, beyond the limits of national jurisdiction. The book reads well. It is recommended for anyone interested in the strategic and international implications of development of the oceans.

337 Furon, Raymond. **The Problem of Water, A World Study**. New York: American Elsevier, 1967, 208p.

This book was written by a member of the Faculty of Science in Paris. He emphasizes the importance of water to life and addresses the amount of water available to sustain it. His calculations suggest the amount of water available can only support 20 billion people, a global population that is expected to be reached by the year 2100. The book is divided into several chapters. The first few discuss the types of water: fresh and salt. The next series of chapters describe the interaction of water with agriculture and industry. Drinking water and water pollution are also discussed. The last half of the text is concerned with development and the search for water.

338 Goldberg, Edward D. **Strategies for Marine Pollution Monitoring**. New York: John Wiley/Interscience, 1976, 310p.

This text discusses the implementation of monitoring techniques, emphasizes the importance of maintaining high quality standards during collection and analysis of samples, and describes methods of analysis of different types of chemicals, such as inorganic and organic species. The treatment of types of analyses and sampling sites is extensive. Additional topics include transport paths in the ocean and petroleum hydrocarbons in marine waters. The text can be technical.

339 Goudie, Andrew, "The Human Impact on the Waters," in Goudie, Andrew, **The Human Impact on the Natural Environment**. Cambridge, MA: MIT Press, 1986, p. 143-190. ****S**

This chapter in a basic text for an introductory course in environmental studies and land resource planning discusses how people have affected the water on our planet. The author, a professor of geography at Oxford University, writes clearly about changing the course of rivers, reclaiming land from the sea, polluting groundwater by chemicals used in agriculture, logging the world's forests to increase erosion and sediment in waterways, and urbanizing rural areas. Numerous figures are included. This is a scholarly and useful source of information for the general reader.

340 Grabler, Raymond, and the Editors of Consumer Reports Books. **Is Your Water Safe to Drink?** Mount Vernon, NY: Consumers Union, 1988, 390p. ****F**

This is a thorough treatment of the problem of drinking water in the United States. The author, a biophysicist, discusses topics such as inorganic, organic, and microbial contaminants in drinking water, the health hazards of drinking water, chlorination, the politics of drinking water, and determining the quality of drinking water. A separate chapter describes bottled waters. Water purification and filtering systems are rated. Illustrations and extensive notes accompany the text.

341 Gullion, Edmund A. (ed.). **Uses of the Seas**. Englewood Cliffs, NJ: Prentice-Hall, 1968, 202p.

This volume describes advances in oceanography in the 1960s and their implications for national and international law. The book was written initially for the American Assembly, a nonpartisan educational organization, for its meeting in May 1968. Topics include: ocean science and marine resources; changing marine law; the Soviet turn to the sea; and an American strategy for the oceans. The focus is on the strategic value of the sea but other questions such as how much can the oceans aid less-developed countries to get food, water, and energy are addressed. Conservation and the effects of pollution are mentioned briefly. The figures and tables are useful. The book is well written, but outdated.

342 Harr, M. E. **Groundwater and Seepage**. New York: McGraw-Hill, 1962, 315p.

This text provides the engineer with a reference for the quantitative solutions to problems of seepage. The mathematical concepts, however, are introduced gradually and explained qualitatively. A background that includes college-level mathematics and some knowledge of civil engineering is strongly recommended. Topics include the fundamentals of groundwater flow, confined flow, unconfined flow, and seepage. Two appendices define specific mathematical techniques necessary to solve flow problems.

343 Hart, Matthew. "Invasion of the Zebra Mussels," in *The Atlantic Monthly*. v. 266 July, 1990, p. 81-87.

This study is a good example of how harmless organisms can reek havoc on the environment. The Zebra mussel was introduced accidentally into Lake St. Clair, near Detroit, in the mid-1980s, by a ship that had taken on ballast water in Europe and discharged it at its North American destination. In August 1988, scientists counted the mussels at 10 sites in Lake St. Clair and discovered densities as high as 200 mussels per square meter. Because the mussels eat phytoplankton, which are the principal food for zooplankton, fish that feed on zooplankton could suffer. This is a startling account of the swiftness by which some species can flourish and take over an ecosystem.

344 Herfindahl, Orris C., and Allen V. Kneese. "Water Pollution," in Herfindahl, Orris C., and Allen V. Kneese, **Quality of the Environment: An Economic Approach to Some Problems in Using Land, Water, and Air**. Washington, DC: Resources for the Future, 1965, p. 10-23.

This is a chapter in a book that stresses the relationship between economic markets and environmental problems. It is a good introduction to water pollutants. Pollutants are broken down into degradable, nondegradable, and persistent. The link between water quality and health also is explored. Three areas of further research are suggested: estimating costs of the deterioration of water quality; devising a system of quality control; and establishing institutional procedures for the management of water quality. The reference list is at the end of the chapter. This work is easy to read and is recommended to all levels of students.

345 Holt, S. J. "The Food Resource of the Ocean," in *Scientific American*, **Man and the Ecosphere**. San Francisco: W. H. Freeman, 1971, p. 84-96.

Written by a member of the Food and Agriculture Organization of the United Nations, this article first appeared in *Scientific American* in September 1969. The author discusses the major fishing areas of the world's oceans and the largest catches of individual fish, oyster, and other marine species. Fishing operations and regulations also are summarized. The problems of exploitation and overfishing are addressed and the consequences are enumerated. Many maps, figures, and photographs are included. This is an excellent and well-written summary of food resources of the ocean.

346 Hovland, M., and A. G. Judd. **Seabed Pockmarks and Seepages: Impact on Geology, Biology, and the Marine Environment**. Lancaster, England: Kluwer Academic Publishers, 1988, 300p.

This text is abundantly illustrated with over 250 figures, maps, and tables. Seabed pockmarks and seepages are produced by both deep and shallow gas leaking through sediments in the ocean floor. The book concentrates on the natural phenomenon, but also explores the harmful side of pockmarks that surface when oil rigs drill through shallow gas pockets. In one incident off the Mississippi delta, such a blowout occurred and the gas ignited. The rig collapsed and eventually sank, leaving a crater in the seabed greater than 600 meters in diameter and 30 meters deep.

347 International Association of Hydrological Sciences. **Groundwater Pollution**. Washington, DC: International Association of Hydrological Sciences, Publication no. 103, 1975, 240p.

The volume is a compilation of papers presented at a symposium in Moscow in August 1971. Papers are grouped into three sections: 1) general studies, which emphasizes transport and sampling of pollutants; 2) natural pollution, which describes the accumulation of salt in groundwater and soil; and 3) artificial pollution, which covers every type of toxic additive, including mineralized industrial waste, municipal wastewater, and commercial fertilizer. The book is technical and a chemistry background is essential. Illustrations are numerous.

348 International Association of Hydrological Sciences. **Hydrogeology in the Service of Man, Volume 1, Keynote Papers**. Washington, DC: International Association of Hydrological Sciences, Publication no. 154, 1985, 301p.

This volume is the first in a four-part series produced from a symposium in Cambridge, United Kingdom, in September 1985. This part is the most general and is aimed at the nonspecialist. Papers cover a wide variety of topics, including discussions of groundwater in different parts of the world, the impact of domestic, industrial, and radioactive wastes on groundwater quality, transport of contaminants, and groundwater conservation. The book is well illustrated and the text is relatively free of jargon.

349 International Association of Hydrological Sciences. **Hydrogeology in the Service of Man, Volume 2, Economic and Social Influence**. Washington, DC: International Association of Hydrological Sciences, Publication no. 154, 1985, 216p.

This volume records the proceedings of a symposium held in Cambridge, United Kingdom, in September 1985 and is the second in a four-part series. Papers discuss the complex interplay between groundwater resources and the development of agricultural and industrial bases. The emphasis is on the impact groundwater has on people in areas of limited supply, such as oceanic islands, Africa, and the Middle East. Little is said concerning the human role in polluting the environment. Although the volume is technical, it is a useful reference. Citations and illustrations are numerous.

350 International Association of Hydrological Sciences. **Hydrogeology in the Service of Man, Volume 3, Groundwater Quality Management**. Washington, DC: International Association of Hydrological Sciences, Publication no. 154, 1985, 207p.

This is a technical volume that is the third in a four-part series published by the International Association of Hydrological Sciences as a result of a symposium in Cambridge, United Kingdom, in September 1985. Papers discuss a variety of topics from a strategy for groundwater quality management to the effects of glaciation. One paper addresses groundwater pollution from landfills. Each contribution contains numerous illustrations and references. A background in chemistry is encouraged.

351 International Association of Hydrological Sciences. **Hydrogeology in the Service of Man, Volume 4, Energy Sources and Groundwater Control.** Washington, DC: International Association of Hydrological Sciences, Publication no. 154, 1985, 243p.

This volume contains papers presented at the 18th Congress of the International Association of Hydrological Sciences in Cambridge, United Kingdom, in September 1985 and is the fourth in a four-part series emphasizing the importance of hydrology to people. Papers discuss details of groundwater studies, such as storage in aquifers, chemical reactions in groundwater, non-Darcy flow, and changes in the levels of aquifers. The volume is technical and knowledge of chemistry is essential. Many illustrations and tables accompany the text.

352 International Association of Hydrological Sciences. **The Influence of Man on the Hydrological Regime with Special Reference to Representative and Experimental Basins.** Washington, DC: International Association of Hydrological Sciences, Publication no. 130, 1980, 453p.

This volume is a collection of papers presented at a symposium in Helsinki, Finland, in June 1980. The nearly 70 papers are organized into two topics, each of which is subdivided further. Subheadings include: effects of urbanization, of agriculture, of engineering, and, finally, of man, on water quality. The last section is particularly interesting. Papers discuss farming, housing, forest drainage, and clear-cutting. The concentration of the symposium was on basins, or wide areas that have the same drainage, and the last part of the volume addresses various aspects of basins. The text is technical and a background in chemistry is suggested.

353 Jones, G. P. **Improvements of Methods of Long Term Prediction of Variations in Groundwater Resources and Regimes Due to Human Activity.** Yarmouth, England: Galliard, 1982, 343p.

This volume is a collection of papers presented at a symposium in the United Kingdom in July 1982 organized by the International Commission on Groundwater and UNESCO. The papers are grouped into four areas: 1) methods of prediction of groundwater resources; 2) effects of human activities on groundwater resources; 3) the sensitivity of parameters concerned with groundwater measurements; and 4) regional studies. The second group of papers is the most interesting and the least technical. Most of these papers discuss the effects of urbanization. All papers have numerous figures, graphs, and tables. A knowledge of chemistry is advisable.

354 Kauffmann, John M. **Flow East: A Look at Our North Atlantic Rivers.** New York: McGraw-Hill, 1973, 284p.

The author attempts to introduce the reader to the scenic rivers of the northeastern United States through the written word. He concentrates on painting a picture of the river in its most pristine setting, such that the contrast to the river in its most developed state is vivid. The book was written in 1973 when conservation action for natural wonders was strong. Rivers were seen largely as commodities to be exploited for their industrial use. The chapter on scenic rivers is particularly good at describing the controversy that surrounded the declaration of the National Wild and Scenic Rivers Act to which no eastern river was named initially. The book is great reading for anyone interested in rivers in general and in the northeast United States in particular.

355 Kenworthy, E. W. "San Diego Cleans Up Once-Dirty Bay as Model for U.S.," in Woods, Barbara (ed.), **Eco-Solutions**. Cambridge, MA: Schenkman, 1972, p. 143-148.

The example of successful pollution control in San Diego, California, is described. Voters initially refused to support the bond issues necessary to finance the construction of treatment plants. Once the bay became too polluted for swimming and the water was considered harmful, the voters approved a bond issue almost three times greater than the earlier one. The cost was approximately $1.50 per household for a total of $60 million with no federal subsidies. The system began operating in 1965 and results were immediate.

356 Kenworthy, E. W. "A Study in Pollution Control: How Seattle Cleaned Up Its Water," in Woods, Barbara (ed.), **Eco-Solutions**. Cambridge, MA: Schenkman, 1972, p. 137-141.

This chapter in a book is a concise discussion of the development of the centralized metropolitan sewage system in Seattle, Washington, and its effect on Puget Sound in the 1950s and 1960s. The growth of algae from abundant phosphorous and nitrogen in Lake Washington also is described. Visibility in the lake declined from 12 feet to 2-1/2 feet in the late 1950s. The establishment of the centralized metropolitan sewage system in 1958 reversed this trend and allowed beaches to be reopened in the city.

357 King, Jonathon. **Troubled Water**. Emmaus, PA: Rodale Press, 1985, 235p.

The author, an investigative reporter, chronicles the chemical contamination of America's drinking water by military and industry. Statistics cited include: one-fourth of underground petroleum storage tanks leak; nearly nine-tenths of toxic waste facilities lie above underground water supplies; and 10 billion gallons of sewage, radioactive waste, and chemicals are injected into the ground each year. The book tells us what is in our drinking water, if filters or bottled water is better, and how the contamination was allowed to get so pervasive. The appendix lists state groundwater agencies. This is a thought-provoking book that will convince many people to stop drinking tap water.

358 Kruse, Allen V. "Approaches to Regional Water Quality Management," in Woods, Barbara, **Eco-Solutions**. Cambridge, MA: Schenkman, 1972, p. 87-136.

The author analyzes the rationale for viewing water quality management from the regional standpoint. He outlines several examples of the river basin as a regional administrative entity, such as the Ruhr Valley in Germany, the English River authorities, the French River basin agencies, and the Delaware River basin. Each geographical area is broken into three sections: 1) situation and accomplishments; 2) historical background; and 3) the administrative and organizational arrangements. Several important criteria are recognized as important to water quality management, including the ability to implement alternatives and the opportunity for affected parties to influence decision-making. The figures are excellent.

359 Lawrence, C. R. and R. J. Hughes (eds.). **Proceedings of the Groundwater Pollution Conference**. Canberra, Australia: Australian Government Publishing Service, 1981, 544p.

The volume is a compilation of papers presented at the first national conference on groundwater pollution in Australia's history. The meeting was in Perth in February 1979. The first three articles are the most general. They cover the contamination of groundwater by uranium mining by-products, the transport of solutes in groundwater, and the absorption of salt into the soil. The majority of the volume, however, discusses specific problems of groundwater pollution in Australia. Some papers are very technical, requiring a strong background in chemistry and mathematics. The first three papers are suitable for anyone interested in the subject.

360 Laycock, George. **Water Pollution**. New York: Grosset & Dunlap, 1972, 69p. ****O**

This is an introductory text that covers most aspects of water pollution. Subjects are the hydrologic cycle, sewage, industrial and agricultural pollution, soil erosion, coal mines, oil, and mercury. The last chapter discusses cleanup of our waterways. The text is simplistic and outdated. This is recommended to younger readers who only want a very general treatment of water pollution.

360a Lee, Douglas B. "Tragedy in Alaska Waters," *National Geographic*, v. 176, n. 2, August 1989, p. 260-263.

This is a good, concise article about the *Exxon Valdez* oil spill in Prince William Sound, Alaska, on March 24, 1989. A team from *National Geographic* arrived the day after the spill to monitor the destruction and cleanup efforts. This is their first report. Good photographs accompany the text.

361 Lowe-McConnell, R. H. **Man-Made Lakes**. London: Academic Press, 1966, 219p.

The volume is a compilation of papers given at a symposium held at the Royal Geographical Society in London in the fall of 1965. Sections in the book describe the history of human-made lakes in the tropics, problems that grew from making lakes in the tropics, the history of human-made lakes in the temperate zone, and problems that arose from creating human-made lakes in the temperate zones. The two parts discussing problems are very informative. The book is easy to read. With the exception of the frontispiece (a map of human-made lakes in Africa), illustrations are not very useful. Because books on this subject are few, this is recommended reading for anyone interested in the topic.

362 MacLeish, William H. **Oil and Water: The Struggle for the Georges Bank**. Boston: Atlantic Monthly Press, 1985, 304p.

The author details the conflicts surrounding the drilling for oil on the Georges Bank, a prime fishing ground off Cape Cod, in 1981. Would fishing forever suffer from the constant oil leaks on the seafloor? Would oil spills threaten the inviolate beaches of Cape Cod? The situation resolved itself when no oil companies bid on leases in the area in 1984. This is an interesting and personal account, emphasizing interviews and dialogues with the people involved.

363 Marx, Wesley. **The Frail Ocean**. New York: Coward-McCann, 1967, 248p. ****O**

The author contemplates the fragility of the world's oceans and the drastic effect that humans have had upon the waters. Topics include: red tides; ocean dumping; overfishing; and pollution of estuaries. Military use of the oceans also is outlined. The book is very descriptive and unfolds like a novel, making it easy to read.

364 Marx, Wesley. **The Oceans**. San Francisco: Sierra Club Books, 1981, 332p.

The book discusses the opportunities that careful, conscientious management of the world's oceans will bring, including the cultivation of fish as food, the erection of floating cities, and recreation. The author begins by outlining the history of abuse of the ocean. Problems cited are overfishing, oil spills, sewage disposal, beach erosion, and marsh draining. Pollution of the ocean has long been condoned by the belief that "dilution is the solution." The author concludes by proposing that societies change their practices on land to safeguard the oceans. References are organized by chapter, which makes finding sources relevant to a particular topic easier to find. This is a well-written treatment that is suitable for everyone.

365 Maurits la Riviere, J. W. "Threats to the World's Water," in *Scientific American*, v. 261, n. 9, September 1989, p. 80-94.

Fresh water represents only 0.01 percent of the Earth's water. The author argues that population growth and poor agricultural practices combine to threaten that water supply. Examples of water pollution are cited, including sewage in San Francisco Bay and chemical effluents in the Rhine River. Freshwater habitats such as rivers, lakes, and coastal zones are discussed. Groundwater contamination is also described. The figures complement the article well. An illustration of the global water cycle is particularly useful. This article is recommended to students of all levels.

366 McCuen, Gary E. **Protecting Water Quality**. Hudson, WI: Gary E. McCuen Publications, 1986, 180p.

This book is part of the Ideas in Conflict series in which various aspects of single issues are debated. Essays by several different authors are grouped into four chapters: 1) "The American Water Crisis: An Overview"; 2) "Blueprints for Clean Water: Ideas in Conflict"; 3) "Groundwater Contamination"; and 4) "Protecting Surface Water." A section on developing reasoning skills through discussion concludes the book. A bibliography of government reports and an index are included. This is an excellent introduction to different ideas for the protection of water quality.

367 Mowat, Farley. "The Siberians," in Woods, Barbara (ed.), **Eco-Solutions**. Cambridge, MA: Schenkman, 1972, p. 149-153.

The chapter is an excerpt from the book with the same title written in 1970. This account of the destruction of Lake Baikal in the former Soviet Union, the largest freshwater lake in the world, illustrates the interaction among logging, erosion, and pollution in lakes and rivers. The story has a happy ending. When the government in Moscow announced plans for a gigantic cellulose and wood-chemical combine on the south shores of the lake in 1962, Soviet citizens cried out and the plant closed. In 1967, the Baikal region was made into a national park, and by the early 1970s, the lake was recovering rapidly.

368 Organisation for Economic Co-operation and Development. **Water Pollution by Fertilizers and Pesticides.** Paris: OECD, 1986, 144p.

This report was compiled by a panel of environmental scientists in the hopes that the member governments of the Organisation for Economic Co-operation and Development would enact restrictions on the use of fertilizers and pesticides. The book has two parts: 1) fertilizers and animal wastes; and 2) pesticides. Topics include: nitrate leaching; nitrate and phosphorus pollution; health issues related to fertilizers and animal wastes; preventive measures to reduce nitrate pollution; pesticides in the aquatic environment; pesticide regulation; pesticide production and use; management measures; future developments; and policy. The text contains many facts. This is a good resource for someone looking for information on this specific subject.

369 Pavoni, Joseph L. (ed.). **Handbook of Water Quality Management Planning.** New York: Van Nostrand Reinhold, 1977, 419p.

This is a hands-on approach to the management of water quality and to all aspects of the planning process. Fifteen authors contributed papers on topics such as the history of water quality management planning in the United States, evaluation of wastewater treatment alternatives, public participation in water quality management planning, and environmental assessments in water quality management planning. The text is somewhat dry but is accompanied by many graphs, tables, and figures that clarify points. This is a practical guide that is recommended to individuals with backgrounds in mathematics or science.

370 Perry, R. (ed.). **Effects of Waste Disposal on Groundwater and Surface Water.** Yarmouth, Great Britain: Galliard, International Association of Hydrological Sciences Publication no. 139, 1982, 278p.

The volume is a compilation of papers presented at a symposium in July 1982 organized by the International Association of Hydrological Sciences and UNESCO. The emphasis is on the effects of waste disposal and chemical pollution on water quality. Papers range from theoretical studies of groundwater flow to specific problems encountered in certain places. Most papers address the transfer of pollutants from landfills to the groundwater supply. Each paper has numerous illustrations and references. Most citations are for academic journal articles. The volume is for the specialist but can be useful to the general reader who has a good background in inorganic chemistry.

371 Pillsbury, Arthur F. "The Salinity of Rivers," in *Scientific American*, v. 245, n. 4, July 1985, p. 54-65.

Written by an engineer, this article discusses the salts in rivers, normally transported to the ocean, that are deposited in the soil when waterways are diverted for irrigation. The process of leaching salts from the ground is described. The Imperial Valley and the Salton Sea of southern California are the examples used. An elaborate plan for controlling and distributing water in the western United States is depicted in a figure.

372 Powers, Charles F., and Andrew Robertson. "The Aging Great Lakes," in *Scientific American*, **Man and the Ecosphere**. San Francisco: W. H. Freeman, 1971, p. 146-154.

This article was first published in *Scientific American* in November 1966. The prehistoric formation of the Great Lakes and their evolution are described and illustrated with excellent figures. The emphasis, however, is on the rapid decay and aging of the lakes caused by humans. One example used to show the acceleration of decay is the building of the Welland Canal, which bypasses Niagara Falls. The canal allowed the sea lamprey of the Atlantic Ocean to migrate beyond Lake Ontario into the other lakes where it killed nearly all the lake trout and other smaller fish species.

372a President's Panel on Oil Spills. **The Oil Spill Problem**. Washington, DC: Office of Science and Technology, 1971, 25p.

This report was commissioned by the president of the United States in 1970 following the oil spill at Torrey Canyon, Santa Barbara, in 1969. The summary of findings and recommendations is particularly insightful. The first point states that the nation is not doing enough to prepare for oil spills such that damage can be minimized. Sections of the report define an oil spill, outline the strategy for combatting a large spill, and discuss avoiding oil spills. The pamphlet is a good overview of oil spills and the resulting effects on the environment.

373 President's Science Advisory Committee, Report of the Panel on Oceanography. **Effective Use of the Sea**. Washington, DC: U. S. Government Printing Office, 1966, 144p.

This report outlines the status of U. S. ocean exploration and research in 1966 and defines the priorities for the future. Topics discussed include food from the sea, modification of the ocean environment, undersea technology, education, and federal organization. Little is mentioned about pollution except to note that the oceans cannot be viewed merely as waste dumps and that the capacity for the reservoirs to carry garbage is not infinite as evidenced by contamination of the Great Lakes, once believed to be capable of disposing of any sewage dumped into them. The publication is severely outdated. It is, however, interesting to read for an historical perspective.

374 Ragone, Stephen (ed.). **Regional Characterization of Water Quality**. Yarmouth, England: Galliard, International Association of Hydrological Sciences Publication no. 182, 1989, 319p.

The volume is a compilation of papers presented at a symposium in Baltimore, Maryland, in May 1989 as part of the annual spring meeting of the

American Geophysical Union. Papers are divided among seven sections, four of which are of interest: lake water quality; effect of agriculture on surface water quality; effect of urban influences on surface water quality; and effect of atmospheric chemistry on surface water quality. Although the text is somewhat technical and full of jargon, the general reader may find the reference useful because the illustrations are numerous. A background in physical sciences, however, would be helpful.

375 Reisner, Marc. **Cadillac Desert**. New York: Viking Penguin, 1986, 582p.

The author tells the history of exploration and development of the West from the perspective of the quest to control water. He begins with a description of the philosophy of cooperative irrigation promoted by John Wesley Powell in the late nineteenth century and continues with an account of the aggressive tactics used by Los Angeles to grab water rights hundreds of miles from the city. (The latter tale was popularized in the movie *Chinatown*.) The majority of the book, however, focuses on the ascension of the Bureau of Reclamation, designed to regulate water in the West and brought to power during the administration of Franklin Roosevelt in the 1930s and 1940s. The stories of greed and manipulation are outrageous. The water crisis of today and the bleak outlook for the future also are discussed. Extensive bibliography, notes, and index accompany the text.

376 Revelle, Roger. "Water," in *Scientific American*, **Man and the Ecosphere**. San Francisco: W. H. Freeman, 1971, p. 57-67.

This is a superb introduction to the importance of water in human societies, written by a distinguished scientist who served on countless government panels, on the effect of increased population on natural resources. The author discusses the obstacles in the way of formulating an intelligent plan for the use and maintenance of the world's water resources. Possible solutions also are described. An excellent figure summarizes the hydrologic cycle for the United States. Errors in water management, such as leaky canals in Pakistan, are cited. The article contains numerous illustrations, all of which are informative.

377 Rodda, J. C., and N. C. Matalas. **Water for the Future: Hydrology in Perspective**. Yarmouth, England: Galliard, International Association of Hydrological Sciences Publication no. 164, 1987, 517p.

This volume contains the proceedings of the symposium on Water for the Future in Rome, Italy, in April 1987. The papers address changes in the study and techniques of hydrology from methods used by the Aztecs and the ancient Hindus to state-of-the-art tools such as expert systems. The volume is useful and informative to the reader who wants to understand the limitations of current hydrologic modeling. A few of the papers include mathematical terms. Most, however, do not. The volume, therefore, is accessible to any interested reader.

378 Simon, Anne W. **Neptune's Revenge: The Ocean of Tomorrow**. New York: Franklin Watts, 1984, 222p. ****O**

The author, a strong proponent of the environment, has written an eloquent and, often alarming, account of the state of the ocean. She criticizes the U.S. government for not having an ocean policy and for retreating from the

Law of the Sea. She urges people to start thinking for the long term, not the immediate future. Despite the deterioration of the ocean that is described, time still exists to implement measures to combat the pollution. This book is a call-to-arms for citizens to get involved. The bibliography is extensive and sources are grouped by chapter.

379 Skjei, Eric, and M. Donald Whorton. "Hazards in the Water," in Skjei, Eric, and M. Donald Whorton, **Of Mice and Molecules**. New York: Dial Press, 1983, p. 121-160. ****O**

A chapter in a book on technology and human survival, this is a good reference for basic descriptions of pollutants in our water. Nonindustrial pollutants that are not limited to modern societies include coliform bacteria, excess nutrients, nitrates, and sediments. Acid rain and toxic substances generated by industry are discussed separately. Mercury, chlorine, and TCE (trichloroethylene) are specific examples. The Clean Water Act and the Safe Drinking Water Act are outlined. In addition, the author suggests home remedies for purifying tap water.

379a Spencer, Page. **White Silk and Black Tar: A Journal of the Alaska Oil Spill**. Minneapolis, MN: Bergamot Books, 1990, 181p.

This is a very personal account of the *Exxon Valdez* oil spill in which 10 million gallons of oil poured into Prince William Sound, Alaska, from a disabled supertanker. The author is an ecologist with the National Park Service in Anchorage and was on her honeymoon when the disaster occurred. After working on the beaches for two months trying to implement monitoring and tracking programs for the oil, she returned home exhausted and wrote an eloquent, impassioned journal about her experiences.

379b Steinhart, Carol E., and John S. Steinhart. **Blowout – A Case Study of the Santa Barbara Oil Spill**. North Scituate, MA: Duxbury Press, 1972. 138p.

This book, written as a balanced account of the Santa Barbara oil spill in 1969 in which 30 million gallons were released, attempts to discuss all of the political, ecological, and scientific ramifications of the accident. Chapters I, V, VI, and VII describe the events in the sequence that they occurred. Chapters II, III, and IV present historical, scientific, and technical information that is necessary for understanding events following the blowout, such as techniques and hazards of drilling for oil. The ecological consequences of the spill are discussed in chapters VII and IX. Legal problems are covered in chapter X. Chapter XI outlines the implications of America's addiction to energy and is, perhaps, the most subjective. The book is fascinating reading and is recommended to everyone.

380 Steinhorn, Ilana, and Joel R. Gat. "The Dead Sea," in *Scientific American*, v. 249, n. 4, October 1983, p. 102-119.

The authors describe the "turnover" of water in the Dead Sea, the world's saltiest lake. For hundreds of years, fresh water that flowed into the lake, which has no outflow, mixed little with the denser, saltier water of the lake but stayed in thin layers at the top of the water column. After irrigation diverted the fresh water entering the lake in the 1960s, the water at the surface became saltier and denser, eventually sinking to the bottom, allowing old water to rise to the surface. This is an interesting article on some unexpected consequences of irrigation.

381 Stewart, John Cary. **Drinking Water Hazards**. Hiram, OH: Envirographics, 1990, 323p.

The author, an environmental chemist and consultant, has written a step-by-step guide to sampling your drinking water, choosing the proper laboratory for its analysis, interpreting the results, and correcting any problems. The types of pollutants found in drinking water are described. A list of references, a bibliography, a glossary, and an index are included. The appendices contain much specific data on testing laboratories and manufacturers of water-related products.

382 Trager, Oliver. "Air and Water Pollution," in Trager, Oliver, **Our Poisoned Planet: Can We Save It?** New York: Facts on File, 1989, p. 2-41. ****O**

This chapter on air and water pollution in a book that contains editorials from American and Canadian newspapers discusses the Monongahela River oil spill, the Clean Water Bill, and the River Protection Bill. A wide variety of newspapers and viewpoints are represented. Each clipping is in the original typeface of the newspaper in which it appeared and is accompanied by the date it was published. The entries are in a loose chronological order. This unusual format makes for very interesting reading.

383 UNESCO. **Discharge of Selected Rivers of the World, Volume I**. Paris: UNESCO, 1971, 70p.

The first in a three-part series, this volume describes the characteristics of stations chosen to participate in the global study of selected rivers. Characteristics of the 1,000 stations include name and number of identification, location, period of operation, and the average monthly and annual discharges for the longest period for which records are available. The data are grouped by continent using the classification scheme proposed by the United Nations: Africa; North America; South America; Asia; Europe; and Oceania. Individual countries are listed alphabetically according to their names in English. The index lists countries by continent in English, French, Spanish, and Russian.

384 UNESCO. **Discharge of Selected Rivers of the World, Volume II**. Paris: UNESCO, 1971, 194p.

The second in a three-part series, this volume contains the monthly and annual discharges recorded at selected stations along certain rivers up to 1964. Data are collected in tables that are self-explanatory. This information is essential to understanding the hydrological cycle and, therefore, to determining sound water policy. The hope of the UNESCO-sponsored study is to gather enough data to decipher trends in discharge fluctuations on a continental or global scale.

385 UNESCO. **Discharge of Selected Rivers of the World, Volume III**. Paris: UNESCO, 1971, 98p.

The third volume in a three-part series, this work catalogues the mean monthly and extreme discharges of selected rivers from all over the world between 1965 and 1969. The book consists primarily of tables of numbers. The information, although a bit dry, is useful to anyone interested in cataloguing variations in discharge through time, assessing concentrations of toxic materials in the rivers, or evaluating the contributions of certain rivers to the oceans. The text is straightforward.

386 United Nations Interregional Meeting of International River Organizations. **Experiences in the Development and Management of International River and Lake Basins**. New York: United Nations, 1983, 424p.

The volume is the summary of the proceedings of the United Nations Interregional Meeting of International River Organizations. The meeting's goals were threefold: 1) the authority of existing international river organizations; 2) the progress among countries for the development and conservation of shared rivers; and 3) the participation of countries in cooperative projects. The volume is divided into three parts. The first details the report of the meeting, the second contains the background papers, which were the basis for discussion, and the third presents papers submitted during the meeting. The material is dry and difficult to follow. The volume, however, provides a realistic view of the bureaucracy and processes surrounding environmental issues on a global scale.

387 U.S. Commission on Marine Science, Engineering, and Resources. **Panel Reports: Science and Environment, Volume 1**. Washington, DC: U.S. Government Printing Office, 1969, 245p.

This first volume in a three-volume set of reports was issued by the U.S. Commission on Marine Science, Engineering, and Resources in 1969 to create an overall plan for a national oceanographic program. The volume begins with an introduction to marine science that primarily addresses its financial health. The next chapter describes basic science and its applications to oceanography. The chapter entitled "Basic Science—Key to Action" is the most interesting. It describes how people affect the shape of the coast, thereby tampering with natural processes and depleting oceanic wildlife. Pollution and environmental monitoring are also highlighted. The remainder of the volume discusses federal and institutional needs and organization. Figures, photographs, and tables are numerous.

388 U.S. Commission on Marine Science, Engineering, and Resources. **Panel Reports: Industry and Technology—Keys to Oceanic Development, Volume 2**. Washington, DC: U.S. Government Printing Office, 1969, 257p.

As the title suggests, this second volume in a three-part series prepared by the U.S. Commission on Marine Science, Engineering, and Resources discusses the effects of industry, private investment, marine engineering, and technology on the oceans. The focus is on accelerating the development of marine resources. A summary of industrial activities in the oceans as of 1969 also is presented. The last chapter gives a summary of ocean industries, including petroleum, natural gas, mining, fishing, aquaculture, and transportation. This is a good source for anyone interested in U.S. government philosophy and policy toward the development of the oceans.

389 U.S. Commission on Marine Science, Engineering, and Resources. **Panel Reports: Marine Resources and Legal-Political Arrangements for Their Development, Volume 3**. Washington, DC: U.S. Government Printing Office, 1969, 211p.

This is the third volume in a three-part series of reports issued in 1969 by the U.S. Commission on Marine Science, Engineering, and Resources. This volume contains six sections, each of which describes a different type of marine resource. The six are fisheries, minerals, oil and gas, fresh water, direct power, and recreation. Regulation, pollution control, and current levels of activity are discussed for each type of resource. Although waste disposal and storage are major uses of the sea, they are not discussed in the report. The volume contains numerous graphs and tables.

390 Wagner, Richard. "Water Pollution: Dilution Is No Solution," in Wagner, Richard, **Environment and Man**. New York: W. W. Norton, 1978, p. 373-407. ****O**

This chapter provides an excellent overview of problems in the environment caused by people. It focuses on the old saying that "dilution is the solution" when it comes to dumping waste into water and argues that dilution is not the answer for densely populated areas. Different types of pollutants are described. An interesting addition is the discussion of animal wastes, reminding us that not all contaminants in the water come from humans. The section on eutrophication is valuable. Solutions to the problems are presented at the close of the ocean pollution discussion. Lakes and thermal pollution also are covered.

391 Wenk, Edward, Jr. **The Politics of the Ocean**. Seattle, WA: University of Washington Press, 1972, 590p.

The politics of the ocean are discussed in terms of the political events of the past and the geographical context of the present (the early 1970s). The first three chapters address policy development, emphasizing the scientific, economic, legal, and social factors that influence it. The next three chapters are studies in coastal management, multinational ocean exploration, the law of the sea, and marine resource development. The last three chapters outline the ability of governments to meet the stresses of future ocean exploitation and present alternative scenarios. The book has insight into the processes involved in determining realistic solutions to international marine exploration and exploitation.

392 Williams, D. I., and D. Anglesea. **Experiments on Water Pollution**. Hove, England: Wayland, 1978, 63p.

This book contains the instructions for numerous simple experiments that can be done to determine the extent of pollution in water. Experiments are grouped into four categories: 1) sampling and analysis of water; 2) purifying polluted waters; 3) effects of polluted water on plant and animal life; and 4) general experiments. Because some of the experiments require special equipment or chemicals, the book is most useful for students. A short glossary and index are included.

393 Zwick, David, and Marcy Benstock. **Water Wasteland**. New York: Grossman, 1971, 494p. ****F**

This book is the report of Ralph Nader's Study Group on Water Pollution and is an analysis of the Water Quality Office of the Environmental Protection Agency and its policies. The report is divided into six parts: 1) "The Danger," which discusses the consequences of water pollution; 2) "Polluters and Protectors"; 3) "Politics, Action and Inaction," which describes pollution in Lake Superior, among other topics; 4) "Law and Order," which critiques legislation; 5) "The New Federalism," which explains subsidies, sanctions, and research sponsored by the federal government; and 6) "Redress," which summarizes the conclusions and suggests recommendations. The appendices contain many lists of environmental agencies. The notes and index are thorough and the work is well researched and thoughtful.

9

Toxic Waste

Images of hypodermic needles on beaches, overflowing garbage in landfills, and foam-covered water in rivers are commonplace in today's polluted world. Rarely is an individual surprised by one more revelation of illegal dumping by either corporations or governments. Regulations are overlooked and the amount of toxic material is staggering. Estimates indicate that the United States produces more than one ton of hazardous waste per person per year—enough to fill the Superdome in New Orleans 15,000 times! Furthermore, the toxic effects of only a few thousand of the 70,000 chemicals used routinely in American industry are well documented.

Hazardous waste—any material that is a health threat—comes in all shapes, sizes, and forms. The common industrial substances include: mercury, lead, and arsenic produced as particles or vapors during copper smelting or the combustion of fossil fuels; asbestos found in insulation, automobile brakes, and mines; petrochemicals manufactured as by-products of the oil-refining process; and vinyl chloride used in plastics. Other materials, most of which are used in daily life, are mothballs, insecticides, battery acid, drain cleaners, bleaches, paints, gasoline, oil, and rat poison.

The problem of hazardous waste is one of the most difficult to remedy and one of the most terrifying. The dangerous substances may be transported hundreds of miles by wind and water. Initial testing may suggest that a substance is not harmful, but several more years of research may indicate serious long-term health risks. Toxic substances may occur anywhere: at home, school, and the office or in food, air, soil, and water. They can percolate slowly into the environment through leaks and improper disposal, or they can poison an area quickly as in the case of an industrial accident. More than half of the hazardous waste produced in the United States is dumped in landfills, which may form the foundation for future housing developments, such as Love Canal, or is injected into wells, which frequently may leak the toxins into the aquifer that supplies a community with groundwater. An additional quarter of the waste is put directly into sewers, rivers, or streams. Only a little over 10 percent of the hazardous material is recycled or processed (see *The Global Ecology Handbook* by Walter H. Corson and the Global Tomorrow Coalition in chapter 16). The problem is so ubiquitous that one estimate suggests nearly 80 percent of Americans live near a waste site and almost 50 percent inhabit counties that contain "dangerous" disposal sites (see reports from the Council on Economic Priorities and the Center for Disease Control).

The sources in this chapter discuss a wide range of topics pertaining to hazardous waste and its disposal. These include: heavy metals, such as lead, cadmium, mercury, and copper (note "Heavy Metal Overload" by James Bellini and "Lead Poisoning" by J. Julian Chisolm, Jr.); solid waste (note "Traditional Environmental Contaminants—Solid Waste" by the Conservation Foundation and *The Solid Waste Dilemma: An Agenda for Action* by the Environmental Protection Agency); PCBs (note *PCBs and the Environment* by John S. Waid); Love Canal (note "A Toxic Ghost Town" and *Laying Waste: The Poisoning of America by Toxic Chemicals* by Michael H. Brown); global dumping (note *Global Dumping Ground* by the Center for Investigative Reporting and Bill Moyers); hazardous waste policy (note *Dimensions of Hazardous Waste Politics and Policy* by Charles E. Davis and James P. Lester and *A Season of Spoils* by Jonathon Lash, Katherine Gillman, and David Sheridan); the contents of landfills (note "Rubbish!" by William L. Rathje); toxic substances and ecological cycles (note "Toxic Substances and Ecological Cycles" by George M. Woodwell) and the link between the Mafia and the disposal of toxic waste (note *Poisoning for Profit: The Mafia and Toxic Waste in America* by Alan A. Block and Frank R. Scarpitti). Excellent overviews are found in "How Dangerous Are Toxic Wastes?" by Julie S. Bach and Lynn Hall, *Hazardous Waste in America* by Samuel S. Epstein, Lester Brown, and Carl Pope, *Who's Poisoning America: Corporate Polluters and Their Victims in the Chemical Age* by Ralph Nader, Ronald Brownstone, and John Richard and *Toxic Threat: How Hazardous Substances Poison Our Lives* by Stephen J. Zipco. More technical treatments include *Toxic Chemicals, Health, and the Environment* by Lester B. Lave and Arthur C. Upton and *Hazardous and Toxic Waste* by Shyamal K. Majumdar and E. Willard Miller. The best sources for information on specific substances are *Toxic Substances Sourcebook* by Steve Ross and Monica Pronen and *Toxic and Hazardous Materials: A Sourcebook and Guide to Information Services* by James K. Webster. Works on pesticides and radioactive material are in chapters 11 and 12, respectively.

394 Bach, Julie S., and Lynn Hall. "How Dangerous Are Toxic Wastes?" in Bach, Julie S., and Lynn Hall, (eds.), **The Environmental Crisis, Opposing Viewpoints**. St. Paul, MN: Greenhaven Press, 1986, p. 171-209. ****O**

This chapter in a book from the Opposing Viewpoints series by Greenhaven Press is organized like a debate with two authors squaring off on some subject. Three topics are discussed: the chemical industry is (not) poisoning America; toxic wastes are (not) causing a cancer epidemic; and the ocean should (not) be used to dump toxic wastes. Ralph Nader and Elizabeth Whelan are two of the authors who participated. Both sides of each issue are heard, making this an extremely interesting and educational book.

395 Barlett, Donald L., and James B. Steele. **Forevermore, Nuclear Waste in America**. New York: W. W. Norton, 1985, 352p.

The authors are investigative reporters at the *Philadelphia Inquirer* and have won numerous national journalism awards. They have written a compelling account of the nuclear waste problem in the United States: each year more and more waste is generated and no one knows where to put it. The book includes discussions on atomic technology, the politics of nuclear power, the reprocessing process for waste, the search for a waste repository, and radiation and national health. Extensive notes, a list of sources, and an index accompany the text. This is a shocking and tragic story that many will find informative.

396 Bellini, James. "Heavy Metal Overload," in Bellini, James, **High Tech Holocaust**. San Francisco: Sierra Club Books, 1986, p. 115-129.

This essay is a chapter in a book that addresses the evils of our modern technological society. Numerous heavy metals are discussed, including cadmium, lead, mercury, and copper. The author cites several alarming statistics and stories, such as one about a family in New Mexico that became seriously ill after consuming pork from a pig that had eaten food fertilized with methyl mercury.

397 Block, Alan A., and Frank R. Scarpitti. **Poisoning for Profit: The Mafia and Toxic Waste in America**. New York: William Morrow, 361p.

Each year, the United States generates over 150 million metric tons of toxic waste, only 10 percent of which is disposed legally. The authors describe where the remaining 90 percent goes. High profits have encouraged the involvement of the Mafia in illegal toxic waste dumping. Contaminants are in the sewers, lakes, garbage dumps, playgrounds, on our highways, and in our national parks. The book is an investigative report that tells the story of numerous detectives and congressional hearings. A horrific story of environmental crime, the book is informative in its descriptions of the toxic waste industry.

398 Brown, Michael. **Laying Waste: The Poisoning of America by Toxic Chemicals**. New York: Pantheon Books, 1979, 351p.

The author was a reporter on the *Niagara Gazette* who, during the mid-1970s, pieced together the story of the contamination around Love Canal. This book is the complete history of Love Canal and the Hooker Chemical Company, from the victims to the perpetrators of the toxic waste dumping. The author also discusses some of the other 50,000 dump sites that the Environmental Protection Agency estimates exist throughout the United States. The book is a well-researched account of a horrifying and urgent problem.

399 Brown, Michael H. "A Toxic Ghost Town," in *The Atlantic Monthly*, v. 263, July 1989, p. 23-28.

This is a brief article in the environment section of *The Atlantic Monthly* magazine that discusses the damage from Love Canal, approximately ten years later. Love Canal, in upstate New York, was the first of America's toxic waste crises and was critical to the establishment of the $10 billion Superfund. During the crisis in 1978 and 1980, 1,030 families were evacuated. Ten years later Love Canal is a 40-acre pile of clay surrounded by a fence and warning signs. A drainage ditch rings the area and takes chemically polluted groundwater to a treatment plant for filtering. The homes are boarded up; some were bulldozed. The cause of Love Canal was the 43.6 million pounds of solvents and pesticide residues dumped in an abandoned canal by the Hooker Chemical Company between 1942 and 1953. The health effects, only now coming to light, are described. The litigation is also discussed.

400 Center for Investigative Reporting and Bill Moyers. **Global Dumping Ground: The International Traffic in Hazardous Waste**. Washington, DC: Seven Locks Press, 1990, 152p.

This eye-opening book is based on a documentary that aired on the Public Broadcasting Service. The reporters discovered the corruption and profit in the underground trafficking in hazardous waste by "following the money." They observed toxics being placed on board cargo ships bound for Third World countries that have less strict regulations on waste disposal than the United States. The authors argue that the environmental laws encourage the generation of toxic waste, much of which is unnecessarily produced. The book is well researched and describes a fascinating and horrible problem intelligently and clearly. The guide to resources is especially informative.

401 Chisolm, J. Julian, Jr. "Lead Poisoning," in *Scientific American*, **Cities: Their Origin, Growth, and Human Impact**. San Francisco: W. H. Freeman, 1973, p. 123-131.

This article first appeared in *Scientific American* in February 1971. Lead is a highly toxic substance that is concentrated in the immediate environment. Because of their high population density, urban areas are the most prone to problems with lead poisoning. The danger is particularly acute for children who live in dilapidated buildings where paint is peeling from the walls. The author discusses the biological effects of ingesting too much lead. The article is scholarly and informative. Several tables and figures are included.

402 The Conservation Foundation. "America's Waste: Managing for Risk Reduction," in The Conservation Foundation, **State of the Government: A View Toward the Nineties**. Washington, DC: The Conservation Foundation, 1987, p. 407-478.

This chapter is part of a report by the Conservation Foundation designed to allow decisions to be made about environmental policy and issues in the coming decade. The discussion begins with a look at the types and amounts of waste in the air, in the water, and on land. Additional topics are: hazard assessment; the regulation and management of waste, including reduction, recycling, treatment, and dispersal; and the types of waste management systems. The conclusions are summarized at the end of the chapter. An extensive reference list, suggestions for further reading, and numerous figures, tables, and graphs accompany the scholarly text.

403 The Conservation Foundation. "Traditional Environmental Contaminants—Solid Waste," in The Conservation Foundation, **State of the Government: A View Toward the Nineties**. Washington, DC: The Conservation Foundation, 1987, p. 107-115.

This section in a chapter of a book written by the Conservation Foundation addresses the generation and the management of waste. Topics include municipal and industrial wastes, mining, agriculture and silviculture, sludges, and other pollution control by-products. The introduction tries to define "solid waste" because no formal environmental definition exists. This is an excellent overview of the waste problem in the United States.

404 The Conservation Foundation. "Toxic and Hazardous Pollutants," in The Conservation Foundation, **State of the Government: A View Toward the Nineties**. Washington, DC: The Conservation Foundation, 1987, p. 135-194.

This well-researched chapter in the report on the state of the environment by the Conservation Foundation is an excellent summary of toxic substances in the world around us. Subjects discussed are: testing and regulation; monitoring programs; organic chemicals such as pesticides and PCBs; inorganic chemicals such as lead, cadmium, and asbestos; regulatory dilemmas; generation, management, and cleanup of hazardous waste; and nuclear waste. The text is dominated by statistics and is accompanied by numerous figures, tables, and graphs. An extensive list of references concludes the chapter.

405 Davis, Charles E., and James P. Lester (eds.). **Dimensions of Hazardous Waste Politics and Policy**. New York: Greenwood Press, 1988, 256p.

This volume is a collection of essays by noted experts in the field of hazardous waste management. Both political and technical issues surrounding the siting of hazardous waste disposal facilities are debated. The policies of various states are summarized. An interesting discussion of the phenomenon of NIMBY (not in my backyard) is included.

406 de Grazia, Alfred. **A Cloud Over Bhopal**. Bombay, India: Popular Prakashan Private, 1985, p. 145.

This is an account of the causes and consequences of the release of a toxic gas cloud by Union Carbide in December 1984, which subsequently spread over the city of Bhopal, India, killing thousands of people. The role of the press is discussed in detail. Much of the book focuses on the negligence of Union Carbide and the assignation of blame for the purposes of litigation. The descriptions of people's reactions to the cloud and the composition of the toxic gas are of interest to the student of environmental problems. The author concludes by suggesting that "a manual of safety" be prepared in every language so that each citizen can read it.

407 Environmental Protection Agency. **The Solid Waste Dilemma: An Agenda for Action**. EPA/530-SW-89-019, Washington, DC: U.S. Government Printing Office, February 1989, 70p.

In this report, the Environmental Protection Agency outlines its goal to manage 25 percent of the nation's municipal waste through "source reduction" and recycling and summarizes the approach it will use to achieve this goal. It will rely heavily on landfills and combustors. The risks associated with these will be evaluated and reduced. The EPA advocates composting of yard waste as one method to decrease the amount of material going into landfills.

408 Epstein, Samuel S., Lester Brown, and Carl Pope. **Hazardous Waste in America**. San Francisco: Sierra Club Books, 1982, 593p. ****O**

This is a thorough investigation of the problem of hazardous waste in America. Topics include dumping in rural America, dumping waste oil, midnight dumping, legislation and regulation of hazardous wastes, groundwater contamination, the technology of disposal, and possible solutions to the problem. Suggestions for citizen action also are included. The book has 14 appendices that contain information such as a list of 8,000 known dump sites in the United States, rankings of states and industries by the amount of waste generated, sources and composition of nonradioactive hazardous wastes, effects of Love Canal, federal statutes on toxic wastes, and a bibliography on hazardous waste exchange technology.

409 Goldman, Benjamin A., James A. Hulme, and Cameron Johnson. **Hazardous Waste Management, Reducing the Risk**. Washington, DC: Island Press, 1986, 316p.

The authors are members of the Council on Economic Priorities (CEP), a public, nonprofit, nonaligned service research organization. The Council has discovered that no systematic critique of hazardous waste companies was done by the Environmental Protection Agency to prepare for disbursement of money from the Superfund for waste disposal. The CEP chose to provide the information necessary for policy makers, regulating agencies, communities, and corporations to make a decision about hazardous waste companies and facilities. Figures and a glossary are included.

410 Grinstead, Robert R. “The New Resource,” in Woods, Barbara (ed.), **Eco-Solutions**. Cambridge, MA: Schenkman, 1972, p. 371-389.

The problem of solid waste is a critical one in the United States, long a society that believed in using and disposing of—not recycling—products. In this article, the author cites ways to cope with the disposal of solid waste. He emphasizes large-scale recycling that necessitates changing the public attitude that goods are to be used only once. The analysis of what is in trash is valuable. Other topics include: recycling methods that were available in the 1970s; prospects for the future; and new technologies that may be required. The example of Mt. Trashmore, a hill used for sledding in Evanston, Illinois, is described.

411 Johansson, Bertram B. “Whisking the Garbage,” in Woods, Barbara (ed.), **Eco-Solutions**. Cambridge, MA: Schenkman, 1972, p. 391-397.

The author describes a system for waste collection and disposal used in the suburbs of Stockholm, Sweden. Pipes are built into apartment buildings. Garbage is dropped into chutes that connect to a vacuum tube that sends the garbage to a centralized collection center. At the facility, the waste is sorted and incinerated. The heat from the incineration is used to heat the apartment buildings. A figure illustrating the system is helpful. The Swedish system was the prototype for a similar system installed in Walt Disney World in Florida.

412 Lash, Jonathon, Katherine Gillman, and David Sheridan. **A Season of Spoils**. New York: Pantheon Books, 1984, 385p.

This book is a well-researched and documented account of the Reagan administration's policies toward the environment. Written in 1984, the work only considers President Reagan's first term. During his tenure, the Environmental Protection Agency (EPA) remained the same size as it was during the Carter administration despite the additional responsibilities the agency had to cleanup and regulate toxic substances and hazardous wastes. As a result, the EPA had a tremendous backlog of work. Oil drilling and national park preservation also are discussed. The book is shocking. Extensive notes and an index are included.

413 Lave, Lester B., and Arthur C. Upton (eds.). **Toxic Chemicals, Health, and the Environment**. Baltimore, MD: Johns Hopkins University Press, 1987, 304p. **S

This is a scholarly and objective collection of valuable papers by noted experts on toxic chemicals and their effects on health and the environment. The perspective of the volume is that because total elimination of the chemicals is not feasible, sensible management is the only recourse. Subjects addressed include: controlling toxic chemicals in the environment; the food chain as a source of toxic chemical exposure; cleanup of contaminated sites; and regulating toxic chemicals in the environment. There are numerous figures and a good index. References are included at the end of each chapter.

414 Majumdar, Shyamal K., and E. Willard Miller (eds.). **Hazardous and Toxic Waste**. Easton, PA: Pennsylvania Academy of Science, 1984, 442p. **S

This collection of papers, edited by two university professors, discusses the technology, management, and health effects of hazardous and toxic wastes. The book is the first of a three-volume series. The papers are grouped in five parts: 1) waste types, treatment, and disposal technology; 2) distribution, selection, and geologic considerations; 3) transportation, emergency response, and preparations; 4) management, regulations, and economic considerations; and 5) environmental and health effects. Numerous figures accompany the text. Some papers are technical. The volume is suitable for a college-level student with some background in science.

415 Melosi, Martin V. **Garbage in the Cities: Refuse, Reform and the Environment, 1880-1980**. College Station, TX: Texas A&M University Press, 1981, 268p.

This is a history of public perception of "garbage as a problem" in cities. The author traces the root to the 1880s when cities were characterized by "mounds of putrefying garbage." Also cited as environmental problems were crowded tenement districts, disease, polluted waters, noise, congestion, and the black smoke that billowed from the newly constructed factories.

416 Montague, Katherine, and Peter Montague. **Mercury**. San Francisco: Sierra Club Books, 1971, 158p.

The book describes the mercury problem that received considerable attention in the late 1960s and early 1970s. The first chapter summarizes the properties of mercury, including those that make it so deadly. The second chapter discusses the mercury poisoning of a town in Japan that first led scientists and doctors to conclude the chemical was highly toxic. The remaining eight chapters cover a variety of topics: fungicides made with mercury; brain damage from mercury poisoning; the companies that manufacture and discharge mercury; and federal regulation. The appendices contain useful information such as the industrial uses of mercury and a state-by-state pollution survey, which now is hopelessly out-of-date. The bibliography is good but is limited to references published before 1971.

417 Nader, Ralph, Ronald Brownstone, and John Richard. **Who's Poisoning America: Corporate Polluters and Their Victims in the Chemical Age**. San Francisco: Sierra Club Books, 1982, 458p. ****O**

This collection of papers written by numerous authors describes environmental disasters caused by American corporations. The emphasis is on the dumping of toxic waste. Crises include Love Canal, the release of PCBs in the Hudson River, and ketone poisoning in Hopewell, Virginia, in 1975. The stories are horrifying both in terms of human interest and corporate accountability. The last paper, "We Are Not Helpless" by Ralph Nader, discusses how citizens can organize to defend the environment and protect themselves. The book is alarming, yet tries to remain objective.

418 Neely, W. Brock, and Gary E. Blau (eds.). **Environmental Exposure from Chemicals**. Boca Raton, FL: CRC Press, 1985, v. 1: 256p., v. 2: 192p.

Scientific experts versed in specific chemicals contributed articles to the volumes. The result is a well-researched and up-to-date resource on the effects of chemicals in the environment. Topics include the transportation, distribution, and fate of chemicals in the groundwater, in the atmosphere, in the air, in the soil, and in the ocean. The text is very technical.

419 *Newsday*. **Rush to Burn: Solving America's Garbage Crisis?** Washington, DC: Island Press, 1989, 269p.

This book was written for citizens and community leaders who wish to understand and to solve the garbage crisis in the United States. Triggered by the garbage barge of 1987 that had no where to go and originally published in *Newsday*, the story concentrates on the throwaway mentality of American society. The book has eight parts, each of which contains a series of vignettes about one subject. Topics include landfills, incineration, illegal dumping, waste management and disposal companies, possible solutions, and recycling. The first section summarizes relevant terminology.

420 O'Leary, Philip R., Patrick W. Walsh, and Robert K. Ham. "Managing Solid Waste," in *Scientific American*, v. 259, n. 6, December 1988, p. 36-45.

Americans generate 140 million metric tons of municipal waste each year and landfill space is getting scarce. If industrial sources are included, the total is 10 billion metric tons of solid waste per year. Two alternatives for waste reduction are conservation and recycling. The development of new types of incinerators for waste management systems is discussed. The article concludes with examples of pollution problems related to landfills such as groundwater contamination and methane production. Possible solutions are outlined. The article is informative, well illustrated, and recommended for everyone.

421 Piasecki, Bruce W., and Gary A. Davis. **America's Future in Toxic Waste Management**. New York: Quorum Books, 1987, 325p.

This collection of papers examines European programs designed to safely dispose of toxic wastes. The authors analyze the programs of 10 European countries from the legal, technical, and planning perspectives, and conclude that a more productive relationship between government and industry is necessary in the United States. Emphasis must shift from dumping to recycling. The volume contains five appendices that discuss the specifics of certain programs, numerous illustrations and tables, an annotated bibliography, and an index. The text is scholarly and readable.

422 Rathje, William L. "Rubbish!" in *The Atlantic Monthly*, v. 264, December 1989, p. 99-109.

The author, an archaeologist who excavates landfills, argues that much of what is said about garbage these days is false: Plastic is a big problem; paper biodegrades quickly in landfills; and modern society generates more trash per capita than ever. His findings are startling and counterintuitive. Foam packaging, for example, composes only 0.33 percent of the volume of landfills whereas paper accounts for 50 percent. Disposable diapers represent 1 percent. In a particularly controversial statement, the author suggests that plastic's resistance to breaking down may be its virtue, as it cannot react with its surroundings to introduce toxins into the environment. This is a fascinating article that should be read by everyone.

423 Regenstein, Lewis. **America, the Poisoned**. Washington, DC: Acropolis Books, 1982, 414p.

Hailed as an important expose, this alarming book tells "how deadly chemicals are destroying our environment, our wildlife, ourselves." The author stresses that when the federal government banned some of the worst chemicals and pesticides, their residues already were found in 99 percent of everyone tested. The toxins, including PCBs, BHC, and DDT, are in air, in water, and in food. Low levels of these chemical compounds were discovered to cause health disorders in laboratory animals. An extensive bibliography, a short list of suggestions for further reading, and an index accompany the text.

424 Ross, Steve, and Monica Pronen (eds.). **Toxic Substances Sourcebook**. New York: Environment Information Center, 1980, 584p. ****F**

Books, journal articles, and professional papers are included in this survey of literature on toxic substances. The editors are thorough. Material from across the spectrum of views on the effect of toxic waste on the environment is incorporated. The result is an excellent, balanced resource for information on toxic substances.

425 Sax, N. Irving. **Dangerous Properties of Industrial Materials, 6th Edition**. New York: Van Nostrand Reinhold, 1984, 3124p.

This is a lengthy tome that serves as a reference for researchers in industrial materials and wastes. The text can be somewhat technical. The work, however, is an invaluable resource on the effects of industry on the environment. It is not recommended as a place to begin studying the topic.

426 Skjei, Eric, and M. Donald Whorton. "Hazardous Wastes," in Skjei, Eric, and M. Donald Whorton, **Of Mice and Molecules**. New York: Dial Press, 1983, p. 218-232.

This chapter in a book on technology and human survival deals concisely with an important problem facing the United States as the twentieth century draws to a close. A brief history of the origins of hazardous dumping is given with reference to Love Canal in upstate New York. Additional topics discussed are the Resource Conservation and Recovery Act, the Superfund, ocean incineration, and the magnitude of the waste problem. Clearly written and informative, this source is recommended to anyone interested in general knowledge about hazardous wastes.

427 Stevens, Barbara J. **Handbook of Municipal Waste Management Systems**. New York: Van Nostrand Reinhold, 1980, 300p.

The author studied seven regions of the United States—New England, New York, the Gulf Coast, the Great Lakes, the mountain states, northern California, and southern California—to identify the costs that affect waste collection from individuals in communities. Five factors were isolated: service; the mix of equipment and labor; transfer and hauling practices; routing and scheduling; and management and organization. The result was recommendations to policy makers for increasing the efficiency and decreasing the cost of waste collection. The appendices describe how the case study sites were chosen and list the forms used to gather the information. This is a fairly dense volume that is recommended to someone who has a keen interest in the subject.

428 Tucker, Richard K., Rick H. Hudson, and M. A. Haegele. **Handbook of Toxicity of Pesticides to Wildlife**. Washington, DC: U.S. Department of the Interior, 1984, 90p.

This is an excellent guide to toxicity levels for animals for the general reader. The entries are listed by the common name of the pesticide. The descriptions are not technical. The small amounts of some substances that produce ill effects are horrifying. The handbook may convince many not to use chemicals on their lawns or in their homes.

429 U.S. Congress, Office of Technology Assessment. **Facing America's Trash: What Next for Municipal Solid Waste**. OTA-O-424, Washington, DC: U.S. Government Printing Office, October 1989, 377p.

This government report, requested by various committees of the House and Senate to help evaluate reauthorization of the Resource Conservation and Recovery Act, discusses a national policy for the prevention and management of municipal solid waste. Topics are policy options, generation and composition of municipal solid waste, prevention of waste, transportation of waste, recycling, incineration, landfilling, and government programs. A glossary and list of acronyms are in appendix A and B, respectively. Numerous tables, graphs, and figures are included. This is a good source of information on a growing problem in our communities.

430 U.S. Environmental Protection Agency. **The Toxics Release Inventory: A National Perspective**. Washington, DC: U.S. Government Printing Office, June 1989, 344p.

This volume describes in great detail the 1987 toxics release inventory (TRI). The chemicals and facilities are discussed. Additional topics are air emissions; discharges to water, on-site land releases and underground injection, off-site transfers of chemical waste, waste treatment, and waste minimization. The volume is full of tables, charts, and maps. This is a useful source for specific data for the nonspecialist that can understand government jargon.

431 Wagner, Richard. "Solid Wastes: Middenheap into Mountain," in Wagner, Richard, **Environment and Man**. New York: W. W. Norton, 1978, 591p.

This is a chapter in a book that is a good general introduction to problems in the environment caused by humans. The chapter describes the different types of solid wastes found in modern society such as paper, cans, bottles, automobiles, tires, and plastics. Disposal techniques are discussed, including landfill, incineration, composting, and recycling. Suggestions for further reading are at the end of the chapter. Scholarly and readable, the chapter is recommended to anyone who wants a qualitative introduction to solid wastes.

432 Waid, John S. (ed.). **PCBs and the Environment**. Boca Raton, FL: CRC Press, 1986-1987 (3 volumes).

This is an ongoing project that is the definitive work on the effects of PCBs on the environment. The volumes consist of reviews of current information by noted experts in the field combined with many tables of data. Some case studies are included. The text is technical and some knowledge of chemistry is recommended.

433 Webster, James K. (ed.). **Toxic and Hazardous Materials: A Sourcebook and Guide to Information Sources**. Westport, CT: Greenwood Press, 1987, 431p. ****F**

This volume contains 11 chapters, each written by a different author, that describe over 1,600 sources of information on toxic and hazardous substances. Most of the entries are annotated to allow the reader to gain insight into the information source. Chapters vary in type of information cited (presumably a reflection of the individual author). This is a good starting point for an investigation of toxic materials.

434 Woodwell, George M. "Toxic Substances and Ecological Cycles," in *Scientific American*, **Man and the Ecosphere**. San Francisco: W. H. Freeman, 1971, p. 128-135.

This article was first published in *Scientific American* magazine in March 1967. The author attempts to dispel the "myth of dilution"—the belief that pouring a little poison into a lot of ocean, air, or soil is harmless. The inaccuracy of this view is shown by the concentration of pesticides in a Long Island estuary. Studies done by the U.S. Atomic Energy Commission on radioactive isotopes in the environment also are summarized. The conclusions were radical for the time: water and wind move toxic substances all over the Earth and animals may concentrate toxic chemicals at very high levels. The figures are excellent.

435 Wright, J. Ward. **Managing Hazardous Waste: A Programmatic Approach**. Lexington, KY: Council of State Governments, Center for the Environment and Natural Resources, State Government Institute, 1986, 84p.

This book is a manual for state, local, and private industry officials to evaluate hazardous waste cleanup. Four case studies of hazardous waste disposal are cited. The exact locations and names of companies and individuals are not specified. A list of problems encountered during the disposal of waste and recommendations for their solution is included. The text is useful only for community or government-sponsored action.

436 Zipco, Stephen J. **Toxic Threat: How Hazardous Substances Poison Our Lives**. New York: Julian Messner, 1986, 193p. ****O**

This is a guide to the hazardous substances in our environment for high school students. The author states that of 70,000 synthetic substances used today, 20 percent are toxic. Fifteen chapters discuss topics such as synthetic organic chemicals, hazardous substances in surface and groundwater, pesticides, herbicides, dioxin, PCBs, radioactive waste, and acid rain. Lists of sources and environmental magazines are included as well as an index.

10

Energy and Resource Depletion

Everything we do requires energy, from driving cars to heating houses, from working on computers to cooking meals, from watching televisions to treating illnesses. Maintaining adequate supplies is critical to "national security." Simply stated, without energy, human society cannot survive. The necessary and logical thing to do, therefore, is to consume the resource intelligently and equitably with as few ill effects as possible. In our modern civilization, however, energy is used inefficiently and most is provided by coal and oil, which produce air pollution and contribute to global warming and acid precipitation. In addition, people in the industrialized nations consume 12 times the energy per capita than do the inhabitants of Third World countries. As the underdeveloped nations grow and modernize, demands for energy will increase, taxing the world's ability to provide the resource.

Energy is derived from a number of sources, including fossil fuels—so termed because they originated as plant and animal matter millions of years ago—nuclear power, electricity, and renewable resources—hydropower, geothermal energy, solar energy, wind power, biomass, photovoltaics, and ocean energy—which can last indefinitely and not harm the environment if managed properly. Most of the world, however, relies on fossil fuels whose consumption has outpaced population growth by nearly four to one since the early 1900s. The United States, for example, gets nearly 90 percent of its energy from fossil fuels, over 40 percent from oil alone.

Throughout history, the solution to greater demand for energy has been to expand the supply as populations burgeon and standards of living improve. In the modern era of concern for the environment coupled with the realities of dwindling oil, gas, and coal reserves and skyrocketing price tags for nuclear power plants, the answer may not be to increase the supply of traditional forms of energy but to turn toward renewable resources. Wind, biomass, geothermal, and solar energies are technologies currently being investigated seriously for future commercial use.

Energy and resource depletion is a vast and complex subject. Clearly, a review of all works on the topic warrants a bibliography unto itself. This chapter merely seeks to provide an assortment of informative books and articles that discuss the basics of energy use in our modern society. Good overviews are "Energy" by the Conservation Foundation, *Energy and Ecology* by David M. Gates, *The Energy Crisis, Opposing Viewpoints* by Bruno Leone and Judy Smith, and *Energy: A Guide Book* by Janet Ramage. A humorous and entertaining way to look at the energy problem is presented in *Energy Unbound* by Hunter L. Lovins, Amory B. Lovins, and Seth Zuckerman. An excellent reference to examine for sources of information on energy is *Energy—Sources of Print and Nonprint Materials* edited by Maureen Crowley. Other topics include: alternative and renewable energy sources (note Ronald M. Benrey and Robert F. Schultz, John O. Blackburn, Denis Hayes); energy consumption in the Third World (note Markus Fritz); energy policy (note Congressional Quarterly and Peter F. Cowhey); energy economics (note Duane Chapman); oil mining (note Richard A. Dick and Sheldon P. Wimpfen); deep Earth gas (note Thomas Gold); synfuels (note Jonathon Lash and Laura King); coal (note Francois Leydet, Michael Parfit, and Richard E. Balzhiser and Kurt E. Yeager); offshore drilling (note Dwight Holing); and oil (note Robert R. Wheeler and Maurine Whited).

437 Balzhiser, Richard E., and Kurt E. Yeager. "Coal-fired Power Plants for the Future," in *Scientific American*, v. 257, n. 3, September 1987, p. 100-107.

The authors describe advances in coal combustion technology that allow the generation of electricity more cleanly and cheaply. A modern plant's pollution control system makes up as much as 40 percent of the cost of the plant but reduces toxic emissions dramatically. In 1900, only 10 percent of wastes were controlled. In 1987, the figure was near 50 percent. This is an informative and well-illustrated introduction for the general reader.

438 Bellini, James. "Living with the Energy Equation," in Bellini, James, **High Tech Holocaust**. San Francisco: Sierra Club Books, 1986, p. 39-58.

This chapter in a book about our polluted Earth argues against the use of fossil fuels and places blame on the energy industry. The author cites the increasing number of cars on the globe as the primary cause of air pollution problems. Carbon monoxide and lead poisoning are discussed in detail. Accidents involving gas tank explosions also are described. This is an interesting and well-written book that has a heavy bias against technology.

439 Benrey, Ronald M., and Robert F. Schultz. **Alternative Energy Sources**. Southfield, MI: Thomas Alva Edison Foundation, 1981, 32p.

This short, entertaining book describes simple experiments one can do at home. The alternative energy sources cited are solar energy, wind energy, ocean thermal energy conversion, tidal energy, energy from trash, coal conversion, geothermal energy, and the fuel cell. Only tidal energy does not have an accompanying experiment. Each experiment discussion begins with a list of things needed. Items are usually readily available. The instructions are straightforward. Although the book is written for the junior high school student, many adults will find it informative.

440 Blackburn, John O. **The Renewable Energy Alternative**. Durham, NC: Duke University Press, 1987, 201p.

The author is an economist and energy planner who believes that renewable energy sources are practical and are the viable alternatives to petroleum-based fuels. The book analyzes current energy use in the United States and outlines conservation measures and the political issues surrounding a transition from oil and gas to renewable fuels. Coal, synfuels, and nuclear power also are discussed. This is a scholarly work that points out solutions to problems as well as the problems themselves. The notes and bibliography are excellent. Numerous tables and figures accompany the text.

441 Brower, Michael. **Cool Energy: The Renewable Solution to Global Warming**. Cambridge, MA: Union of Concerned Scientists, 1990, 89p.

This report by the Union of Concerned Scientists is a concise treatment of an important environmental problem: the link between energy consumption and global warming. Chapters include fossil fuels and global warming, energy strategies, solar energy, wind energy, biomass energy, hydroelectric power, and the path to a renewable future. Numerous figures, tables, and a list of references accompany the text. This straightforward book is recommended to anyone who wishes a broad overview of energy.

442 Brown, Harrison. "Human Materials Production as a Process in the Biosphere," in *Scientific American*, **The Biosphere**. San Francisco: W. H. Freeman, 1970, p. 115-124 or in *Scientific American*, **Man and the Ecosphere**. San Francisco: W. H. Freeman, 1971, p. 106-114.

The article, written by a former governmental advisor and professor of geochemistry at the California Institute of Technology, first appeared in *Scientific American* in September 1970. The emphasis is on the interrelationships among energy, metals, and prosperity. The author describes the use and extraction of products such as metals and concrete, stressing that they are nonrenewable resources. Unfortunately, these products are seldom recycled and frequently become solid waste that requires disposal. The author further illustrates the problem of finite natural resources in a world of increased population and development using the example of projected demand for steel in the early twenty-first century. The historical background of metal use also is included. Well-written and illustrated, the article is recommended to anyone interested in the relationship among mining, manufacturing, solid waste disposal, and recycling.

443 Chapman, Duane. **Energy Resources and Energy Corporations**. Ithaca, NY: Cornell University Press, 1983, 365p.

This book on energy economics covers most aspects of the energy industry. Topics discussed are economic theory; competition, monopoly, and growth; energy use; petroleum, natural gas, and oil companies; coal, nuclear power, and utilities; and problems in public policy. The author analyzes the social and environmental costs of nuclear power and coal as fuels in preparation for addressing public policy issues. The book is particularly informative in its detailed descriptions of the corporate structure in each energy industry. Recommendations for lowering energy consumption conclude the book. The glossary and index are helpful. The text was written for an undergraduate course in energy economics but can be used as a reference by anyone.

444 Clark, Joel P., and Merton C. Flemings. "Advanced Materials and the Economy," in *Scientific American*, v. 255, n. 4, October 1986, p. 50-57.

This article describes the new materials being created such as advanced metals, ceramics, polymers, and composites. The authors emphasize the properties of the materials that will revolutionize modern industry. The environmental and health effects of manufacturing and using the new compounds, most of which are unknown, are mentioned briefly.

445 Congressional Quarterly. **Energy Policy, 2d Edition**. Washington, DC: Congressional Quarterly, 1981, 274p. **F**

Energy supply and cost in the last two decades of the twentieth century are discussed. It examines the 1973 Arab oil embargo and the resultant feeling of the American public that reliance on foreign oil was too high. Alternatives to foreign imports are described, including nuclear power, solar energy, and synthetic fuels. The conflict between energy and the environment also is detailed. The major energy speeches of Presidents Nixon and Carter, energy legislation from 1973-1980, a glossary, and a bibliography conclude the volume. This is an excellent source of facts on energy use and supply in the United States during the 1970s.

446 The Conservation Foundation. "Natural Resources: Land, Water, and Energy—Energy," in The Conservation Foundation, **State of the Government: A View Toward the Nineties**. Washington, DC: The Conservation Foundation, 1987, p. 237-256. **O, S**

This section in a chapter of a report by the Conservation Foundation designed to allow informed decisions to be made about the environment summarizes energy resources. Types of energy are oil, natural gas, coal, electricity, nuclear power, hydroelectric power, biomass, and solar, geothermal, and wind energy. A discussion of international trends in energy consumption and production closes the section. This is a well-researched and scholarly overview.

447 Cook, Earl. "The Flow of Energy in an Industrialized Society," in *Scientific American*, **Energy and Power**. San Francisco: W. H. Freeman, 1971, p. 81-89.

The United States is used as the example of an industrialized society through which the author traces the flow of energy. With only 6 percent of the world's population, the U.S. uses 35 percent of the world's energy. Topics discussed include: the rates of increase of energy use in various sectors of the society such as households, industry, and transportation; efficiency of energy use; and environmental constraints on energy consumption. The author argues that the inability of society to dissipate waste heat from energy use, not scarcity of resources, will severely limit energy consumption. Many figures augment the text. The article is well written and informative.

448 Cowhey, Peter F. **The Problems of Plenty**. Berkeley, CA: University of California Press, 1985, 447p. **S**

Written by a political scientist, this book emphasizes the relationship between energy policy and international politics. Topics include: strategies to manage energy problems; the evolution of energy policy from 1919 to the mid-1980s; and the causes and consequences of international management strategies. The discussion reveals how government and corporate policies toward the management of energy resources led to decreased competition and strong coalitions of either importing or exporting interests. The author suggests changes in the current hierarchy may lead to cooperation between the two sides. The scholarly work is accompanied by extensive notes and a thorough bibliography.

449 Crowley, Maureen (ed.). **Energy: Sources of Print and Nonprint Materials**. New York: Neal-Schuman, 1980, 341p.

This volume lists nearly 800 organizations that provide information on energy. The author hopes that the interested reader either will follow publications put out by the individual agencies of interest or will request to be placed on an agency's mailing list. The book is organized by type of agency such as government, activist, labor, university-affiliated research, and corporation. Each agency is described and relevant publications are mentioned. The indexes are particularly useful.

450 Cuff, David J., and William J. Young. **United States Energy Atlas, 2nd Edition**. New York: Macmillan, 1986, n.p. **F

This atlas is an excellent pictorial representation of the distribution of types of energy sources in the United States. Numerous graphs, maps, and charts are included. The first part of the atlas describes nonrenewable resources and the second part discusses renewable resources. The volume allows the reader to see quickly what areas of the country depend on what type of energy. This is a fascinating and handy reference tool.

451 Deudney, Daniel, and Christopher Flavin. **Renewable Energy**. New York: W. W. Norton, 1983, 431p.

The authors have compiled abundant information about renewable energy sources, capitalizing on advances made in the 1970s and early 1980s. The book begins with an overview of the main energy sources of the present, namely oil, natural gas, coal, and nuclear power. Subsequent chapters discuss alternative types, including solar, ethanol, energy from urban wastes, hydroelectric, wind, geothermal, and wood. The positives and negatives of each are debated. The book concludes with descriptions of institutions necessary for transition to a society based on renewable energy sources and sustainable development. Copious notes and an extensive reference list augment the text.

452 Dick, Richard A. and Sheldon P. Wimpfen. "Oil Mining," in *Scientific American*, v. 243, n. 4, October 1980, p. 182-189.

The process of oil mining, which becomes profitable when prices are high, is described. The technique has been around for two centuries and is practiced most in the modern world in Canada and the U.S.S.R. The author asserts that oil mining is relatively clean when compared with conventional mineral extraction. He does not rule out, however, the possibility of strip-mining. This is an informative article about a method that may be important in the future.

453 Editorial Research Reports. **Energy Issues, New Directions and Goals**. Washington, DC: Congressional Quarterly, 1982, 204p.

This collection of 10 reports was published initially as separate Editorial Research Reports which are sent to libraries and educational institutions on a subscription basis. Topics include America's nuclear waste backlog, the western oil boom, solar energy, wood fuel, wind and water, nuclear fusion, synthetic fuels, and deep sea mining. Many illustrations complement the simple and direct text.

454 Edmonds, Jae, and John M. Reilly. **Global Energy: Assessing the Future**. Oxford, England: Oxford University Press, 1985, 317p.

This book is the result of a research project on forecasting energy use and its impact on global carbon dioxide levels sponsored by the Institute for Energy Analysis. The majority of the book tries to estimate energy demand and supply in the coming decades. Each type of energy is described, including conventional oil, coal, nuclear energy, solar and wind energy, biomass, and hydroelectric energy. From the discussion, the authors conclude that fossil fuels will continue to be dominant. Because these contribute large amounts of carbon dioxide to the Earth's atmosphere, fossil fuels are emphasized in the model for carbon dioxide levels that is explained in the last third of the volume. Numerous tables and figures are used. The text is technical and recommended to the reader with a background in beginning college mathematics.

455 Freeman, S. David. **Energy: The New Era**. New York: Walker, 1974, 386p.

This book is written from the perspective of the mid-1970s, shortly after the oil crises of 1973 and 1974, and, as such, is slightly outdated. Nevertheless, the discussions of energy demands, energy versus the environment, and the politics of energy are still relevant and useful reading. The book is well written and recommended to any interested person. The reader of the 1990s may find the chapter on the next 10 years (through 1984) fascinating. The author warns that the backlash against the environmental movement in the mid-1970s will escalate environmental problems. He argues for stronger controls on tankers to avoid oil spills and advocates conservation.

456 Fritz, Markus. **Future Energy Consumption of the Third World**. Oxford, England: Pergamon Press, 1981, 206p.

The author critiques earlier global models that predicted Third World energy demand and presents his own analysis for energy use in the Third World with specific emphasis on nuclear and electrical power. The purpose of the study was to estimate the future demand for nuclear energy in the underdeveloped nations. The author does not attempt to define nuclear power as good or bad but merely strives to understand the extent to which the Third World will become dependent on this energy source.

457 Gates, David M. **Energy and Ecology**. Sunderland, MA: Sinauer, 1985, 377p. ****O**

This is a comprehensive guide to the ecological and environmental consequences of energy extraction, conversion, and use. The 14 chapters cover topics such as the atmosphere, energy resources, energy demand, solar energy, biomass, coal, carbon dioxide and climate change, acid deposition, and alternative energy sources. Each chapter has an introduction, discussion, and ecological impact analysis. The bibliography is comprehensive. Units and conversion factors are listed in the appendix. This is an excellent overview of numerous energy-related environmental problems.

458 Gever, John, Robert Kaufmann, David Skole, and Charles Vorosmarty. **Beyond Oil: The Threat to Food and Fuel in the Coming Decades**. Cambridge, MA: Ballinger, 1986, 304p.

The authors outline the "carrying capacity" of the Earth in terms of resources. They discuss energy efficiency, farm output, population growth and standards of living, and productivity. Several alarming statements are made, including the exploration for oil will expend more energy than it gains, the return to coal as an energy source will create skies as polluted as during the Industrial Revolution, and energy costs will be so great that farmers will not be able to run their farms. The authors maintain that these situations can be avoided if we mitigate our desires. Numerous figures and graphs accompany the text.

459 Gibbons, John H., Peter D. Blair, and Holly L. Gwin. "Strategies for Energy Use," in *Scientific American*, v. 261, n. 9, September 1989, p. 136-143.

The authors discuss ways to cut down energy use which can increase energy supplies and slow climatic changes. They state that human activity expends the amount of fossil fuel in one year that was produced by nature in one million years. The article begins with descriptions of fossil fuels: oil, natural gas, and coal. It continues with a discussion of alternative energy sources, including solar power, hydroelectric power, nuclear power, biomass, and geothermal production. The concept of efficiency also is introduced. Excellent figures accompany the well-written text. This is a good overview for everyone.

460 Gold, Thomas. **Power from the Earth: Deep Earth Gas—Energy for the Future**. London: J. M. Dent & Sons, 1987, 224p.

The author, a natural scientist, advocates the theory that much methane lies deep below the Earth's surface, having accumulated during the formation of the planet. His argument comes from models of planetary accretion and from observations of the distribution of oil and gas fields, which occur in a variety of geologic settings. In Turkey, for example, the oil fields are in plains, mountains, and in between. Testing done on oil from all the fields suggest it is of common origin. The explanation is that hydrocarbons from deep within the Earth seep upward because they are light and collect in fields all over the world. If Dr. Gold's hypothesis is correct, more than enough methane is below the Earth's surface to fuel our society in the decades to come. The theory, however, is highly controversial.

461 Gray, Charles L., Jr., and Jeffrey A. Alson. "The Case for Methanol," in *Scientific American*, v. 261, n. 9, September 1989, p. 108-114.

The authors, both of whom work at the U. S. Environmental Protection Agency's Emission Control Technology Division, describe the possibilities of various methanol-fueled automobiles. They maintain that only modest changes are required to the automotive and energy industries to introduce a methanol-based transportation system in the United States. Increased efficiency of automobiles and decreased emissions are two of the advantages of methanol described. A table of characteristics of alternative fuels is informative. The figures are excellent. The text is clearly written and easy to follow.

462 Hapgood, Fred. "The Quest for Oil," in *National Geographic*, v. 176, n. 2, August 1989, p. 226-259.

The article traces the activity of exploration geologists as they search for oil throughout the world. The work takes the author from the rainforests of Papua-New Guinea to the frozen tundra of Alaska. The guidelines and tools geologists use to look for oil are described clearly and concisely. Excellent photographs accompany the text.

463 Hayes, Denis. **Rays of Hope**. New York: W. W. Norton, 1977, 240p.

This book discusses the transition from a petroleum-based world economy to one that is either nuclear or solar based. The author argues that a nuclear-powered world will be more centralized and authoritarian than a solar-powered one. He cites as evils of the nuclear world its carcinogenic, mutagenic, and toxic by-products. The book was written in the mid-1970s when fears of oil shortages were at their peak. Looking from the perspective of the early 1990s when oil is cheap and plentiful, the book seems pessimistic and alarming. Its message, however, is relevant. The index is useful for the reader who does not wish to read the entire tome.

464 Holing, Dwight. **Coastal Alert, Ecosystems, Energy, and Offshore Oil Drilling**. Washington, DC: Island Press, 1990, 126p.

This book is the product of the Natural Resources Defense Council and the Central Coast Regional Studies Program. It describes the effects of offshore drilling on the marine environment as well as explaining the drilling and leasing process. Citizen action in California and energy alternatives to offshore drilling also are examined. Sample letters to send members of Congress are included. The glossary is quite useful. This book is an excellent resource for individuals interested in protecting America's coastlines.

465 Hubbert, M. King. "The Energy Resources of the Earth," in *Scientific American*, **Energy and Power**. San Francisco: W. H. Freeman, 1971, p. 29-40.

This article is an introduction to the energy sources on the Earth, including the sun, the tides, the Earth's heat, nuclear fission and fusion fuels, and petroleum. Estimates of the world's coal and oil reserves are from 1971 and are a bit out-of-date. Fission and fusion are explained both in the text and in a figure. The well-written and scholarly article is recommended to anyone interested in world energy reserves.

466 Knoepfel, Heinz. **Energy 2000: An Overview of the World's Energy Resources in Decades to Come**. New York: Gordon & Breach, 1986, 281p.

The author begins by describing energy today and tracing the history of energy as it relates to human evolution. Included in this discussion are such far-reaching topics as acid rain in the greenhouse and Greek philosophies on the structure of the atom. The second half of the book introduces the energy sources of tomorrow—nuclear fission, solar energy, and nuclear fusion. The text is very well written and illustrated. The references are thorough. This book is enthusiastically recommended to the reader interested in learning how energy is created from nuclear and solar processes.

467 Lash, Jonathon, and Laura King (eds.). **The Synfuels Manual**. New York: Natural Resources Defense Council, 1983, 306p.

This manual summarizes the history, the present state and the impact of the use of synthetic fuels. The book is divided into eight chapters: 1) "Introduction"; 2) "History"; 3) "The Technology of Synthetic Fuels Production"; 4) "Environmental Impacts of Coal Conversion"; 5) "Environmental Impacts of Oil Shale Development"; 6) "Socioeconomic Impacts"; 7) "Citizens, Synfuels, and the Environmental Laws"; and 8) "Other Useful Laws." Numerous figures, a short glossary, and extensive notes are included. This is an excellent source.

468 League of Women Voters of California. **The Cost of Electricity**. Sacramento, CA: League of Women Voters, 1986, 35p.

This project was sponsored by the League of Women Voters in 1985 to gather research, disseminate information and educate the public about rising electricity prices in California. Utility bills are explained by outlining the factors that affect the costs the power company must pay such as fuel, planning uncertainties, construction, environmental safeguards, and supply changes. The discussion of the history of electricity use in California is particularly worthwhile. The League suggests that people could easily use 30 percent less electricity without compromising their standard of living. The booklet is clear, concise, and well illustrated.

469 Leone, Bruno, and Judy Smith. **The Energy Crisis, Opposing Viewpoints**. St. Paul, MN: Greenhaven Press, 1981, 143p. ****O**

Part of the Opposing Viewpoints series, this book contains four chapters, each of which debates a central issue: Is there an energy crisis? Is nuclear power an acceptable alternative? Is the energy consumer exploited? and What are our energy alternatives? The pros and cons of each question are argued in either four or six papers. A bibliography of periodical articles concludes each chapter. A section entitled "Discussion Activities" is designed to help the reader learn to think critically. Topics include: distinguishing between fact and opinion; distinguishing between statements that are provable and those that are not; distinguishing primary from secondary sources; understanding stereotypes; and determining priorities. This is an excellent introduction to the central issues surrounding energy in the United States.

470 Leydet, Francois. "Coal vs. Parklands," in *National Geographic*, v. 158, December 1980, p. 776-803.

This article summarizes the debate between environmentalists and energy advocates over the mining of coal, oil shale, and tar sands in southern Utah near Zion, Bryce, Canyonlands, Capitol Reef, and Arches National Parks, Cedar Breaks National Monument, and Glen Canyon National Recreation Area on Lake Powell. The issues are complex. One of the most pressing is the allocation of water, a precious resource in the West. The increase of smog in the Moab Valley also is described. The article explains the controversy well and depicts the magnificent scenery superbly.

471 Lovins, L. Hunter, Amory B. Lovins, and Seth Zuckerman. **Energy Unbound**. San Francisco: Sierra Club Books, 1986, 390p. **O

This "fable for America's future" is about how the "sensible" use of energy will solve a myriad of problems including acid rain, the national debt, Third World poverty, and farm failures. The authors, who are experts on alternative energy sources, tell the story of a midwestern housewife who becomes secretary of energy. Through the experiences of the housewife, the reader discovers the problems of energy supply and cost, nuclear policy, special interest lobbying, and the practicality of alternative energy. The only illustrations are political cartoons. This book provides an entertaining way to learn about energy in the United States.

472 Luten, Daniel B. "The Economic Geography of Energy," in *Scientific American*, **Energy and Power**. San Francisco: W. H. Freeman, 1971, p. 107-117.

This is an introduction to the uses of energy as a function of geography. Topics include worldwide patterns of oil production, refining, shipping and consumption; transportation of crude oil and natural gas within the United States; and per capita energy consumption of different nations including industrialized and nonindustrialized countries. The emphasis is on the rate-limiting steps of the flow of energy resources such as shipping and storage. The many figures illustrate most of the important concepts of the scholarly text. The article is recommended to anyone interested in understanding the complexities of the world's energy industry and the economic interdependence of nations.

473 McGowan, Linda Baine, and John O. M. Bockris. **How to Obtain Abundant Clean Energy**. New York: Plenum Press, 1980, 276p.

The authors begin by asking "What is energy?" They continue by examining various forms of energy and by explaining the physical concepts behind turning different sources into energy. Sunlight is one example. Alternatives to the dominant forms of energy used in modern industrialized nations also are discussed. Other topics are the approaching energy disaster and the hydrogen economy. Many figures and a good index accompany the text. The thorough treatment of the subject makes this book invaluable for anyone interested in the energy industry.

474 National Academy of Sciences. **Resources and Man**. San Francisco: W. H. Freeman, 1969, 127p.

This collection of papers analyzes demand for several important resources as of 1969, including food from both land and sea, minerals from onshore and offshore exploration, and energy. The future abundances of these resources also are discussed. Additional topics include the human ecosystem and population trends in the U.S. and across the globe. This is a very thorough and scholarly treatment by a distinguished panel of experts. It is recommended to any individual interested in the problem of our dwindling resources. The book is intriguing because it was written over 20 years ago.

475 National Research Council Panel on Operational Safety in Marine Mining. **Mining in the Outer Continental Shelf and in the Deep Ocean**. Washington, DC: National Academy of Sciences, 1975, 119p.

This report of a blue-ribbon panel of the National Research Council attempts to determine the potential of resources on the continental shelf (areas of the ocean adjacent to continents and less than 200 meters deep), to evaluate the state of ocean mining, and to identify the environmental problems associated with the mining. In addition, the book considers how to meet the need for trained personnel to do the mining. The panel was most concerned with the technology of ocean mining and its economic viability. Environmental issues are secondary.

476 Oatman, Eric F. (ed.). **Prospects for Energy in America**. New York: H. W. Wilson, 1980, 220p.

Several papers by politicians, government officials, and scientists address the debate over the future of energy in the United States. Contributors include: Jimmy Carter; Howard M. Metzenbaum; Daniel Yergin; and Isaac Asimov. The papers are grouped into seven chapters: 1) "Stalking the Problem"; 2) "Inventing Solutions"; 3) "The Nuclear Dilemma"; 4) "Coal and Natural Gas"; 5) "Alternative Fuels"; 6) "Sun, Air, Earth, and Water"; and 7) "Conservation: Cheapest Fuel of All?" Each chapter has an editor's introduction that summarizes the contents of the papers that follow. Clearly written and nontechnical, this book is an excellent overview of the energy debate in America.

477 Parfit, Michael. **Last Stand at Rosebud Creek: Coal, Power, and People**. New York: E. P. Dutton, 1980, 304p.

The author chronicles the experiences of 18 people who become involved in the effort to stop the construction of two coal-burning, electricity generating plants. Their arguments were that air pollution from the burning coal would be great, strip mining would leave tremendous scars, and the health effects of inhaling coal steam are not adequately documented. The acceptable alternative is nuclear power. The book has an interesting perspective that is based on the pragmatic decisions of people who must live near the plant generating electricity.

478 Perrine, Richard L., and W. G. Ernst (eds.). **Energy for Ourselves and Our Posterity**. Englewood Cliffs, NJ: Prentice-Hall, 1985, 434p.

The papers in this volume were initially presented at a colloquium on the energy crisis at UCLA in early 1980. Most important issues of the supply and demand of energy are covered, including the technical and economic aspects and governmental policies and regulations. Authors include politicians, economists, engineers, and scientists. The focus is on the United States although a few papers discuss other parts of the world. Oil, gas, coal, and nuclear fuel are described thoroughly. One drawback of the volume is the lack of coverage of solar energy and synthetic fuels. Subsequent to the symposium, oil and gas prices fell sharply in the mid-1980s, which makes some of the papers seem prematurely alarming and outdated.

479 Ramage, Janet. **Energy: A Guide Book**. Oxford, England: Oxford University Press, 1983, 345p. ****O**

This book on global energy consumption is divided into three parts: 1) energy today; 2) technologies; and 3) the future. The first part summarizes the global picture and describes the primary forms of energy used in different countries. It also contains an excellent section on the interpretation of data. The second part outlines the methods used to convert both renewable and nonrenewable resources into energy. Examples are: oil and natural gas, coal, nuclear power, energy from water, solar energy, and wind power. The third part discusses the consequences of excessive energy demand, including air pollution, acid rain, biological effects of radioactivity, and the greenhouse effect. Also presented are predictions for alternative futures. The suggestions for further reading are good—both sides of an issue are represented. This is an excellent overview for the individual interested in the energy industry.

480 Sassin, Wolfgang. 'Energy," in *Scientific American*, v. 243, n. 3, September 1980, p. 118-132.

In a special issue of *Scientific American* that features global economic development, this article focuses on energy demands in the coming years. Most increases will come from the less developed countries, requiring a "transfer of technology" from the industrialized nations. Energy efficiency, energy demand, consumption rate, fossil fuel resources, and energy investments are discussed.

481 Siegel, Mark A., Carol D. Foster, and Nancy R. Jacobs. **Energy: An Issue of the 90's**. Wylie, TX: Information Plus, 1989, 124p.

This is part of the information series on current topics. Chapter titles are "An Energy Overview," "How Long Will the Oil Glut Last?," "Natural Gas," "Oil and Gas Reserves," "Coal—A Slow but Steady Growth Industry," "Nuclear Power," "Nuclear Waste," "Alternative Energy Sources," "Electrical Power and Supply," and "Acid Rain." The text is accompanied by many graphs, maps, and illustrations. This is an excellent summary of the state of the energy industry in the late 1980s.

482 Singer, S. Fred. "Human Energy Production as a Process in the Biosphere," in *Scientific American*, **The Biosphere**. San Francisco: W. H. Freeman, 1970, p. 105-114.

The article first appeared in *Scientific American* in September 1970. It discusses the influence people have on the natural cycles of the Earth when they release the energy stored in fossil and nuclear fuels. Aspects of energy consumption by humans described by the author include the release of byproducts such as carbon dioxide and water vapor during the burning of fossil fuels and the discharge of heat by a nuclear power plant. Global warming as a result of increased carbon dioxide content in the atmosphere also is outlined. There is an excellent introductory source on environmental problems related to energy consumption.

483 Smith, Duane A. **Mining America**. Lawrence, KS: University of Kansas Press, 1987, 210p.

This volume is an account of the relationship between the mining industry and the environment from 1800 to 1980. The author chronicles the history of the mining industry's attitudes toward conservation from early exploitation to later protection. In doing so, he describes most forms of ore extraction, including hydraulic mining, placer mining, and open-pit mining. Pollution from smelting and dredging also is discussed. Excellent photographs, extensive notes, a thorough bibliography, and a good index accompany the text. The book is an interesting study of the changes in perspective that occur within an industry as it ages.

484 Starr, Chauncey. "Energy and Power," in *Scientific American*, **Energy and Power**. San Francisco: W. H. Freeman, 1971, p. 1-15.

This article summarizes people's need for and exploitation of energy, particularly in the Western industrialized nations. In the United States, for example, electrical energy demands doubled during most decades of the twentieth century. The consequences of the increase are debated. The sources of U.S. energy also are described. The article concludes with an analysis of the pollution created by various energy sources, such as particulates in the air from the burning of fossil fuels, and a list of possible compromises. The article first appeared in a special issue of *Scientific American* in September 1971. It is well written and suitable for anyone.

485 Summers, Claude M. "The Conversion of Energy," in *Scientific American*, **Energy and Power**. San Francisco: W. H. Freeman, 1971, p. 93-106.

This article was first published in September 1971 in a special issue of *Scientific American*. The author focuses on the efficiency by which any fuel is converted to a useful form, an important concept in an industrial society where high-quality energy is turned into waste heat. Figures are informative and illustrate most of the points of the article, including the efficiency of different energy converters, conversion pathways, mechanisms of different energy producers, and growth curves for energy demand and depletion of resources. This is an excellent source for someone interested in understanding the demands of an industrialized society for energy and the efficiency of energy sources.

486 Thomas Alva Edison Foundation. **Energy Conservation—Experiments You Can Do**. Southfield, MI: Thomas Alva Edison Foundation, 1981, 32p.

This short book describes 11 experiments that can be done at home with a moderate amount of time and effort. Experiments include: measuring the amount of water that drips from a leaky faucet; determining the difference in the amount of water used in a shower or a bath; estimating drafts at windows; and building a solar-powered hot dog cooker. A discussion of various energy sources begins the volume. Also helpful is a paragraph that explains how to read the gas and electric meters. Most people will enjoy this fun book!

487 Tosato, G., J. Brady, P. Essam, M. Finnin, G. Giesen, S. Rath-Nagel, H. Vos, and D. Wilde. **Energy After the Eighties—A Cooperative Study by Countries of the International Energy Agency**, Amsterdam: Elsevier Science, 1984, 298p.

This volume is the product of an international study organized by the International Energy Agency of the Organisation for Economic Co-operation and Development to assess the potential benefits of new technologies in the 14 participating countries. The focus was to develop a quantitative understanding of interrelationships in a complex, global system. The first half of the book outlines the assumptions and basic ingredients for the model of estimating the benefits. The second half discusses the results of the model by country. The book is detailed and quantitative. It is only recommended to people with an interest in energy and a background in mathematics.

488 Udall, Stewart, Charles Conconi, and David Osterhout. **The Energy Balloon**. New York: McGraw-Hill, 1974, 288p.

Published in 1974, this book, designed to be provocative, seems overly alarmist from the perspective of the 1990s. The author, a former secretary of the interior, details the self-deceptions of the United States government and the vested interests of the oil companies that led to the energy crisis of the early 1970s. Chapter titles include "Energy Illusions and Energy Alternatives," "Shaping the Human Environment," and "The Fateful Dilemma: Domestic Consumption and Global Scarcities." Although the book is outdated, much of the advice is still relevant.

489 Wheeler, Robert R., and Maurine Whited. **Oil, From Prospect to Pipeline, Fifth Edition**. Houston, TX: Gulf, 1985, 147p.

This is an excellent introductory overview of what it takes to find, drill, and get oil to market. Seven chapters discuss world oil, finding and drilling for oil, ownership of oil, and the oil's worth. The four appendices are particularly useful. Appendix A lists abbreviations used in oil reports. Appendix B describes stratigraphic terminology, the geological eras, and oil and gas provinces both inside and outside the lower 48 states. Appendix C contains sample leases and deeds, whereas appendix D summarizes energy equivalents. An oil dictionary concludes the book.

490 Yanarella, Ernest J., and Anne-Marie Yanarella. **Energy and the Social Sciences: A Bibliographic Guide to the Literature**. Boulder, CO: Westview Press, 1982, 347p.

This volume is designed to help professionals, students, and librarians familiarize themselves with a broad spectrum of sources on energy. The emphasis is on works from the field of social sciences. The bibliography contains over 4,000 entries, none of which are annotated. Introductory essays begin each major section. This is a comprehensive treatment that is slightly outdated.

11

Pesticides, Cosmetics, Food Additives, and Other Chemicals

We consume hazardous substances every day. "Better living through chemistry"—to quote one well-known advertising slogan—is the attitude of one major corporation and, indeed, that of the industrialized world. People want perfectly shaped produce, nonfattening cake, and wrinkle-removing eye cream. The consequence is that hazardous chemicals are in the food we eat, the cosmetics we use, and the water we drink. Some pesticides, food additives, and cosmetic ingredients are linked to birth defects, cancer, seizures, and nausea. Pesticides are particularly loathsome; agricultural runoff seeps into municipal water supplies and evaporates into the atmosphere. Accidents during the manufacture and transport of the toxic chemicals can expose hundreds, even thousands, of people to the deadly substances. A Worldwatch Institute report published in 1988 estimated that pesticides alone poison between 400,000 and 2 million people each year and that 10,000 to 40,000 of them die. The problem is most serious in Third World countries where farm workers use the toxic chemicals without taking precautions. Illness also occurs, however, in populated areas near major agricultural centers in the developed nations. The town of McFarland in the productive farmland of the San Joaquin Valley of central California, for example, has a cancer rate among children that is significantly higher than the national average. The cancer risk is believed greater for children because they are very susceptible to toxic chemicals and they eat a lot of fruits and vegetables.

The use of pesticides has increased dramatically over the past decades. In 1966, the United States applied 300 million pounds of pesticides. That figure climbed to over 1 billion pounds in 1987 (see *The Global Ecology Handbook* by Walter H. Corson and the Global Tomorrow Coalition in chapter 16). The amount used globally grows by an estimated 12 percent per year. This trend continues despite two findings that suggest pesticides are not as effective as they once were. First, crop losses from insects hovered around 32 percent in the 1940s before the introduction of the toxic chemicals. After their use became routine, losses jumped to 37 percent. Second, fewer than 10 species of insect were resistant to pesticides in 1938. The number of resistant species is now considerably more than 400. The response by the agriculture industry is to apply more and stronger pesticides, which, in turn, produce additional resistant species.

Public outcry and government testing resulted in the banning of some of the more dangerous pesticides in the United States, including DDT and BHC. The federal government suspects, however, that as much as 25 percent of the banned substances are sold overseas by American companies where they are used routinely on fruits and vegetables eventually returned to the United States. Less than 1 percent of the imported food is inspected.

The environmental concern and health consciousness of the 1980s led to the publication of many books on the hazardous substances in the food we eat and the cosmetics we use. Some of the better sources that list the harmful ingredients include: *Guess What's Coming to Dinner* by Americans for Safe Food; *It's All on the Label* by Zena Block; *House Dangerous* by Ellen J. Greenfield; *Harmful Food Additives* by Milton Houben and William Kropf; *Pesticide Alert* by Lawrie Mott and Karen Snyder; and *Everything You Want to Know About Cosmetics* by Toni Stabile. An excellent balanced overview of the benefits and risks of many chemical compounds, including food additives and pesticides, is *The Two Faces of Chemistry* by Luciano Caglioti. Other topics are: hunger and health (note Claude Aubert); the history of battling the use of pesticides and preservatives (note Warren J. Belasco); the pros and cons of pesticide use (note Allen A. Boraiko, CONCERN); alternatives to pesticides (note Terry Gips); DDT (note Thomas R. Dunlap); the accident at the Union Carbide plant in Bhopal, India (note Larry Everest, David Weir); dioxin or agent orange (note Michael Gough, Alastair Hay, Fred Wilcox); pesticides in imported food (note Shelley Hearne); air pollution from pesticides (note Robert E. Lee, Jr.); the pesticide lobby (note Robert Van Den Bosch); herbicides (note Carol Van Strum); NutraSweet (note H. J. Roberts); government testing of food additives (note Jacqueline Verrett and Jean Carper); and the management of parkland (note Sewell E. Simmons).

491 Americans for Safe Food. **Guess What's Coming to Dinner: Contaminants in Our Food**. Washington, DC: Center for Science in the Public Interest, 1987, 56p. ****O, F**

The Center for Science in the Public Interest has produced a thorough overview of the contaminants in our food. The book discusses pesticides, herbicides, antibiotics frequently administered to livestock, food additives including preservatives, artificial sweeteners, and flavor enhancers, and PCBs, some of the most toxic chemicals in our environment. The issues surrounding the use of these substances are also debated. This is a handy reference for individuals interested in knowing what is in the food Americans eat and it is suitable for the general reader.

492 Aubert, Claude. **Hunger and Health**. Emmaus, PA: Rodale Press, 1985.

The author wrote the book to focus attention on the drawbacks of modern food technology and the benefits of autonomous food production, including less pollution and less impact on ecosystems. The text is organized under 11 questions. Examples are: "Can agricultural practices be increased without using chemical fertilizers and pesticides?" "Do chemicals used in agriculture affect health?" and "Can irrigation programs be developed that are not harmful to health?" Suggestions for further reading follow each question. The bibliography is extensive.

493 Belasco, Warren J. **Appetite for Change: How the Counterculture Took on the Fast Food Industry 1966-1988**. New York: Pantheon Books, 1990, 311p.

The history of unprocessed food and vegetarianism in the United States is traced in this book. The author tells how preservatives and pesticides fell from favor as ecological political action became fashionable. Food companies were targeted. The response of the industry was merely to repackage and advertise the same food as "lite" or "all-natural." The enthusiastic reaction of consumers to the "new" products ensured that the food companies had won the battle. The writing is a bit turgid, obfuscating some of the important information.

494 Block, Zena. **It's All on the Label: Understanding Food, Additives, and Nutrition**. Boston: Little, Brown, 1981, 326p. ****O, F**

This is a comprehensive guide to the ingredients in processed food. Each chapter, beginning with chapter 5, goes through one food group, such as dairy products, baked foods, cereals, pasta, meat, frozen foods, and so forth. The earlier chapters are overviews of topics including food laws, additives, safety, how foods are preserved, colors, and flavors. The book is extremely practical. Using it, the consumer can make more informed decisions about food, menus, and diets. This is recommended to everyone curious (or worried) about food contents.

495 Boraiko, Allen A. "The Pesticide Dilemma," in *National Geographic*, v. 157, February 1980, p. 145.

This is a thoroughly researched and objective article accessible to the general reader about pesticides and their consequences. Some facts cited: herbicides reduce erosion and conserve soil moisture by allowing less tillage; insect resistance is accelerating; harvest yields have doubled since 1945; and exposure to some pesticides can alter brain activity for a year. Testing, pesticide residue on crops, and health effects also are discussed.

496 Caglioti, Luciano. **The Two Faces of Chemistry**. Cambridge, MA: MIT Press, 1983, 218p. ****O**

The book, written by a professor of organic chemistry at the University of Rome, is translated from the original Italian version. The author presents a balanced view of the benefits and hazards of modern chemical compounds and their by-products. Subjects include: food additives; fertilizers; pesticides; medications; soaps; plastics; fluorocarbons; leaded gasoline; and cosmetics. The data are primarily from the United States, Western Europe, and Japan. The discussion is objective; the assessment of risk is unbiased. The author conveys the sense of excitement that accompanies a new discovery and the feeling of horror that follows revelation of the side effects of such discoveries. This is a very enlightening book. An excellent index, bibliography, and tables complement the text.

497 Center for Study of Responsive Law. **Eating Clean**. Washington, DC: Center for Study of Responsive Law, 1982.

The collected readings of this manual on diet, food safety, and chemical additives include newspaper articles, government literature, congressional testimony, and political essays. They are divided into five parts: 1) "Filthy Food"; 2) "Additive Avalanche"; 3) "The Pollution Stew"; 4) "Defending and Strengthening the Food Safety Laws"; and 5) "Taking Steps Toward Safer Food." Lists of major grocery chains and food action groups are compiled in the back of the book. The introduction by Ralph Nader is alarming. This is an interesting volume in both content and layout.

498 CONCERN. **Pesticides: A Community Action Guide**. Washington, DC: CONCERN, 1987.

This is an informative overview of pesticide issues that serves as a guide to what citizens can do to alleviate and prevent community health problems caused by pesticides. The discussion of the pros and cons of pesticide use is worthwhile and unbiased. Also described are various action groups that target pesticide use. The approach is not technical and the text is concise. This book is recommended to individuals who want to learn about the pesticide problem in the United States but do not have unlimited time to dedicate to the effort.

499 Dunlap, Thomas R. **DDT: Scientists, Citizens, and Public Policy**. Princeton, NJ: Princeton University Press, 1981, 318p.

The author traces the history of the controversy surrounding DDT, beginning with the development and regulation of pesticides prior to the appearance of DDT. He then describes the discovery of DDT and the growing awareness of its harmful effects on the environment. Public opposition to the pesticide led to the filing of lawsuits in 1967 and the banning of DDT by the Environmental Protection Agency in 1972. The text is straightforward and easy to follow. Technical language is reserved for the appendices, which discuss the chemical structure of DDT, the production of and tolerance level for pesticides, and vapor phase chromatography.

500 Edwards, C. A. (ed.). **Environmental Pollution by Pesticides**. London: Plenum Press, 1973, 542p.

This volume takes a balanced approach between the point of view of ecologists that all pesticides are bad and the belief of agriculturalists that large quantities of pesticides are necessary. Each chapter is written by a different scientist. The introduction describes the historical development of pesticides. The two indexes, general and chemical, describe the techniques for measuring organic contaminants and the errors involved.

501 Edwards, Clive A. **Persistent Pesticides in the Environment**. Cleveland, OH: CRC Press, 1970, 170p.

A short, thorough, and scholarly discussion of common pesticides, this book is part of the CRC (Chemical Rubber Company) Monoscience series, which seeks to provide reference works in a single subject. Topics include: pesticides in the physical environment, such as soil and water; pesticides in the biota; and control of pesticides. Numerous tables and figures augment the text. The bibliography is extensive. The appendix summarizes relevant legislation in 1969 and 1970. Although this volume is outdated, it still is recommended as a useful source of background material.

502 Everest, Larry. **Behind the Poison Cloud**. Chicago: Banner Press, 1985, 192p.

The book is a description of the Bhopal incident in which Union Carbide released from its pesticide factory a cloud of toxic gas over the city of Bhopal, India, on December 3, 1984. The author interviewed numerous people including Union Carbide workers, Indian government officials, scientists, and doctors. He also visited the Bhopal plant and concluded that Union Carbide was directly responsible because the plant's design and location were faulty. A controversy surrounding the suppression of medical treatment for victims also is discussed. This is a thorough, somewhat biased account.

503 Gips, Terry.**Breaking the Pesticide Habit: Alternatives to Twelve Hazardous Pesticides**. Minneapolis, MN: International Alliance for Sustainable Agriculture, 1988, 372p.

The author describes safe alternatives to the pesticides cited by the Pesticide Action Network in their Dirty Dozen campaign. The campaign singled out 12 pesticides that were believed extremely hazardous and responsible for most of the deaths from pesticides recorded throughout the world. They also do considerable damage to the environment by affecting entire ecosystems. This is a very practical guide that is most useful to the grower but is still of interest to the general public.

504 Gough, Michael. **Dioxin, Agent Orange: The Facts**. New York: Plenum Press, 1986, 290p.

This treatment of the hazards of dioxin, also known as Agent Orange, is unbiased and scholarly. The author, an expert on microbiology, is the director of the Risk Science Institute in Washington, D.C., and was formerly in the Division of Health and Life Sciences of the Congressional Office of Technology. The book addresses topics including the carcinogenic properties of dioxin, the short-and long-term effects of exposure to the substance, the possible link to birth defects, and the subject most in the news, the cause of death of Vietnam veterans.

505 Greenfield, Ellen J. **House Dangerous**. New York: Vintage Books, 1987, 234p. ****O, F**

This is a guide to pollutants in the home and office. The second and third chapters are walks through your home and office with stops in every room. Pollutants typical of each area are discussed. The fourth chapter tells how to reduce indoor pollution. Suggestions include using nonphosphate detergents and nonammonia cleaners. The fifth chapter chronicles government regulations on indoor pollution. The sixth chapter lists common household pollutants in alphabetical order. The glossary is indispensable. A checklist for your home and the addresses of manufacturers and distributors of less polluting products are extremely valuable. This is an excellent practical reference for everyone.

506 Hallenbeck, W. H., and K. M. Cunningham-Burns. **Pesticides and Human Health**. New York: Springer-Verlag, 1985, 176p.

The book is a list of pesticides that was prepared as a reference for medical and public health professionals, toxicologists, environmentalists, consumer advocates, lawyers, regulators, industrial hygienists, and users of pesticides. Under each entry the following characteristics appear: the use of the pesticide; acute exposure effects; chronic exposure effects; suspected effects; toxicology; active ingredients; and additional information. The appendices contain a glossary, a bibliography, and compilations of canceled and suspended pesticides. The alphabetical organization of the book makes it easy to use.

507 Hay, Alastair. **The Chemical of Scythe, Lessons of 2,4,5-T, and Dioxin**. Disaster Research in Practice. New York: Plenum Press, 1982, 264p.

The volume concentrates on the problems posed by the use of toxic chemicals. Chapters discuss dioxins, hexachlorophene, trichlorophenol, and chlorain. In addition, case studies of hazardous materials at Love Canal, defoliants in Vietnam, and dioxin in Italy are included. Part of the Disaster Research in Practice series, this volume strives to present thorough, scientific summaries about the most urgent problems for the general reader.

508 Hearne, Shelley. **Harvest of Unknowns: Pesticide Contamination in Imported Foods**. New York: National Resource Defense Council, 1984, 70p.

The author describes pesticides in imported foods and the unique problems these toxic substances pose. The United States typically has stricter controls on the use of pesticides than other countries have, creating a dilemma—Does the United States forego the foods that contain prohibited pesticides or does it allow these goods to enter the domestic market? The question is tricky, particularly because the United States is the chief producer and exporter of many of the pesticides. This is a thought-provoking and informative guide to the complex issues of foreign pesticide use.

509 Herfindahl, Orris C., and Allen V. Kneese. "The Use of Chemicals as Pesticides," in Herfindahl, Orris C., and Allen V. Kneese (eds.), **Quality of the Environment: An Economic Approach to Some Problems in Using Land, Water, and Air**. Washington, DC: Resources for the Future, 1965, p. 35-52.

This is a chapter in a book that balances the economics of environmental problems against the benefits of the standard of living of contemporary society. It is a good overview of pesticides and other chemical substances in our environment. Sections in the chapter include: the effects of pesticides; regulation; the next steps necessary to combat the problem; choices and their implementation; and prospects for easing the problem. A short reference list concludes the chapter.

510 Houben, Milton, and William Kropf, M.D. **Harmful Food Additives**. Port Washington, NY: Ashley Books, 1980, 146p. ****0**

This is an informative book that explains the content of the food we eat. Chapter headings include: "The Twenty Worst Additives in Food"; "The Triple Threat: Excessive Use of Sugar, Salt, and Fat"; "Water and Additives"; "The Toxic Chemicals That Surround Us"; and "Food Trends." A chart in the chapter titled "Selective Approach" analyzes 100 foods for their contents and lists the best brands of each food type. The bibliography and index are short. The book is a good source of information but is outdated.

511 Hunter, Beatrice Trum. **How Safe Is Food in Your Kitchen?** New York: Charles Scribner's Sons, 1981, 88p.

Contrary to what the title implies, this book does not concentrate on pesticides and additives in food, but rather it details hazards associated with cookware, glassware, wooden utensils, storage containers, and food wraps. Microwave ovens, barbecues, and water softeners also are described. A list of sources and an index conclude the book.

512 International Organization of Consumers Unions. **The Pesticide Handbook: Profiles for Action**. Penang, Malaysia: International Organization of Consumers Unions, 1986.

This practical reference lists the health effects of more than 40 pesticides commonly used throughout the world. Also described are important international pesticide issues. One example is the regulation of pesticides that may be banned in one country but used to grow food in another. What happens when the first country imports food from the second? Another example is the release of pesticides into the air, allowing them to spread over much of the globe.

513 Kallet, Arthur, and F. J. Schlink. **100,000,000 Guinea Pigs**. New York: Grosset & Dunlap, 1933, 320p.

This interesting work outlines the "dangers in everyday foods, drugs, and cosmetics," and is noteworthy for its publication date, 1933, which is well in advance of the environmental and consumer activist movements of the 1960s. Examples of toxins include: arsenic and lead on fruits and vegetables; some hair dyes; and some medicines. Most of the items, however, are "fakes"—products that promise to cure an affliction but, in reality, do not. The authors also suggest ways to guard against the dangers. The text is worthwhile from an historical context. It is not particularly useful for modern analysis.

514 Lee, Robert E., Jr. (ed.). **Air Pollution from Pesticides and Agricultural Processes**. Cleveland, OH: CRC Press, 1976, 264p.

This technical volume compiles the known information about the contribution of pesticides and agricultural processes to air pollution. The book has four parts: 1) chemistry and physics of airborne pesticides; 2) toxicity of airborne pesticides; 3) pesticide legislation; and 4) air pollution produced by agricultural procedures other than that generated by pesticides. This is a thorough treatment of the subject that is suitable for the individual with some familiarity with physical sciences.

515 le Riche, W. Harding. **A Chemical Feast**. New York: Facts on File, 1982, 204p.

The author outlines the relationships among chemicals, diet, and disease. Topics are: chemicals and the normal diet; food additives and illness; natural poisons in food; pesticides; food poisoning; water; drugs in pregnancy; cancer and food; allergies; heart disease; and hyperactivity. The point the author makes—that increased processing of food has not brought new health dangers—is well taken. A list of references and an index conclude the book. This is an excellent source of information for both natural and artificial substances in food.

516 Mooney, Booth. **The Hidden Assassins.** Chicago: Follett, 1966, 230p.

This is an analysis of the foods we eat and a critique of the federal government for not imposing stricter regulations. Chapters discuss, in general terms, the problems within the food industry such as the economic benefit to farmers the use of pesticides gives and the harm to consumers the substances cause. One chapter outlines what can be done. Appendices include: truth in packaging and consumer complaints; seizure actions reported by the Food and Drug Administration (FDA) in August 1965; and a debate on whether pesticides are necessary. The book is outdated, as the discussion of saccharin illustrates.

517 Mott, Lawrie. **Pesticide Reregistration: An Evaluation of the EPA's Progress.** San Francisco: National Resource Defense Council, 1986, 117p.

The author studied the efforts of the Environmental Protection Agency to reevaluate the safety of pesticides that were inadequately tested when they were first introduced. The results of the investigation are both alarming and reassuring—early testing was not thorough and additional testing was done where appropriate. The work is scholarly and well researched. This is an intriguing book that is recommended to the reader who is curious about the operations of the government in the area of public safety.

518 Mott, Lawrie, and Martha Broad. **Pesticides in Food: What the Public Needs to Know.** San Francisco: National Resource Defense Council, 1984, 123p.

An informative discussion of the pesticides commonly found in food, the book does an excellent job of sifting through the available data to clarify the issues and dangers for the average citizen. Most people will be disturbed by the lack of government testing and regulation and appalled by the influence of the agricultural lobby. The authors suggest several ways to reform the present system. This is a good overview of pesticides in food and is suitable for anyone.

519 Mott, Lawrie, and Karen Snyder. **Pesticide Alert.** San Francisco: Sierra Club Books, 1987, 128p. ****O**

Written by two members of the Natural Resources Defense Council, this book is a guide to pesticides in fruits and vegetables. Also discussed is the impact of pesticides on the environment. The book is divided into three sections. The first describes pesticide residues found in food. The second lists each type of fruit and vegetable in alphabetical order, identifying the five most common pesticides used to grow each item. The third section summarizes federal regulation of pesticides. Extensive notes, a compilation of additional sources and further reading, a glossary, a bibliography, and an index are included. This is an indispensable reference for anyone concerned about toxins in food.

520 O'Brien, Mary. **On the Trail of a Pesticide: A Guide to Learning about the Chemistry, Effects, and Testing of Pesticides**. Eugene, OR: Northwest Coalition for Alternatives to Pesticides, 1984, 148p.

This is an excellent reference on pesticides that is aimed at the citizen activist. In addition to listing common pesticides, their uses, and health effects, the author describes the impact each pesticide has on the environment. For example, DDT ingestion by birds causes them to produce eggs with abnormally thin shells. The discussion of toxicology is particularly worthwhile. The slant of the book is strongly for the environment, resulting in little debate over the benefits versus the consequences of pesticide use. An exhaustive treatment, the book is recommended to anyone interested in becoming involved.

521 Okun, Mitchell. **Fair Play in the Marketplace**. DeKalb, IL: Northern Illinois University Press, 1986, 345p.

This is an account of the "first battle for pure food and drugs" in the late nineteenth century, which, according to the author, was the precursor of the national consumer movement launched after the Pure Food and Drug Act of 1906. During the late 1800s, modernization of manufacturing led to widespread fear of the introduction of poison during the canning, packaging, and distribution of food and medicine. The author recounts the conflicts between the new industry and the politicians who tried to regulate it. The public health implications also are discussed. The volume has extensive footnotes, a bibliography, and an index. The treatment is thorough but is difficult to follow.

522 Pratt, Christopher J. "Chemical Fertilizers," in *Scientific American*, **Man and the Ecosphere**. San Francisco, W. H. Freeman, 1971, p. 236-246.

This article, describing the types and uses of chemical fertilizers, first appeared in *Scientific American* in June 1965. The author stresses that to obtain higher yields of crops, more and more fertilizer must be applied—the relationship is not linear; three times the fertilizer may be required to double the yield. The illustrations are excellent and include maps of the United States that show nutrient deficiencies; drawings of corn leafs that have certain mineral deficiencies; a table of basic processes used to manufacture fertilizer; and a graph of the availability of nutrients in certain types of soils.

523 Ridgeway, James. **Who Owns the Earth**. New York: P. F. Collier, 1980, 154p.

This book lists household items, including food and drugs, and explains where each originates, how it is produced, what raw materials are used, and how valuable it is. Items are catalogued in general categories, such as food, fertilizers, fish, beverages, luxuries and narcotics, fuels, forest products, fiber, metals, precious metals and stones, and the human body. Discussions are enlightening and entertaining. Photographs accompany most entries. This is an informative book recommended to anyone interested in how we use the world's resources.

524 Roberts, H. J. **Aspartame (NutraSweet), Is It Safe?** Philadelphia: Charles Press, 1990, 315p.

The author is a doctor who wrote the book after seeing numerous patients with symptoms that he could not relate to specific illness but could associate with the use of aspartame (NutraSweet). The symptoms included headaches, memory loss, vision changes, sleep disorders, and nausea. The history of approval of aspartame by the Federal Drug Administration also is summarized. An extensive bibliography and thorough index accompany the text.

525 Rodale, J. I. **Our Poisoned Earth and Sky**. Emmaus, PA: Rodale Books, 1964, 735p.

A dense treatise on the environmental hazards of our planet published over 25 years ago, the book contains five sections, a summary of conclusions, and several appendices. The sections are 1) "The Food on Our Plates"; 2) "Drugs and Cosmetics"; 3) "The Water We Drink"; 4) "The Air We Breathe"; and 5) "Radiation and Fallout." Each is very thorough and further divided into chapters. The appendices list food preservatives, enrichment additives, occupational toxins, and state legislation. Also included is a glossary of atomic terms. The book is an excellent reference but is extremely outdated.

526 Sapolsky, Harvey M. (ed.). **Consuming Fears: The Politics of Product Risks**. New York: Basic Books, 1986, 256p.

An excellent introduction to the realities of product risks, this book tells the histories of six products linked to serious health effects, the common themes that unite them, and the political repercussions of product fears. The products are cigarettes, meat and dairy products believed to cause heart disease, tampons, salt, artificial sweeteners, and insulation containing formaldehyde. Each is described in its own chapter written by an expert in political science, public heath, or consumer issues. The first and last chapters, penned by the editor, give an overview of product safety issues and a summary of the politics surrounding controversies, respectively. Extensive notes and a thorough index accompany the lucid and scholarly text.

527 Simmons, Sewell E. **Parkland Pest Management**. Sacramento, CA: State of California Department of Food and Agriculture, 1985, 109p.

This is a manual for park personnel, owners of small farms, professional gardeners, groundskeepers, and landscapers. It contains seven sections: 1) "Distinguishing Characteristics"; 2) "Pesticide Use"; 3) "Invertebrate Pest Management"; 4) "Vertebrate Pest Management"; 5) "Plant Disease Management"; 6) "Weed Management"; and 7) "Bibliography." The volume is very matter-of-fact and does not dwell on the hazards of pest control substances, although a list of environmental concerns appears below each entry. This is a good reference for pesticides used in public facilities.

528 Skjei, Eric, and M. Donald Whorton. "Hazards in the Food Supply," in, Skjei, Eric, and M. Donald Whorton, **Of Mice and Molecules**. New York: Dial Press, 1983, p. 161-217.

A chapter in a book on technology and human survival, this is a thorough treatment that can stand on its own. In addition to discussing sugar, salt, fiber, and cholesterol, the authors describe food additives such as nitrites, saccharin, and estrogens. Compounds in pesticides and herbicides also are examined, including chlorinated hydrocarbons, organophosphates, carbamates, dioxin, paraquat, and TCDD. Alternatives to insecticides are outlined. The book is informative, well written, and nontechnical.

529 Stabile, Toni. **Everything You Want to Know About Cosmetics; or, What Your Friendly Clerk Didn't Tell You**. New York: Dodd, Mead, 1984, 285p. ****O, F**

Cosmetics, as defined in this book, are all grooming products, including toothpaste, foot powder, lipstick, and eyeliner. These products typically are not regulated by the United States Congress and its agencies, and the companies that produce the items are not required to be federally registered. As a result, cosmetics are either not tested or they are inadequately researched prior to being sold over the counter. The author provides information about grooming products and answers the most commonly asked questions. She also gives suggestions on how to minimize the health risks.

530 Trager, Oliver. "Chemicals, Toxic Waste, Pesticides and Asbestos," in Trager, Oliver, **Our Poisoned Planet: Can We Save It?** New York: Facts on File, 1989, p. 2-41.

This is a chapter in a book that contains editorials from American and Canadian newspapers. Topics include: the Bhopal accident; violations by Union Carbide; an accident on the Rhine River; and barred pesticides. A wide variety of newspapers and viewpoints are represented. Each clipping is in the original typeface of the newspaper in which it appeared and is accompanied by the date it was published. The entries are in a loose chronological order. This unusual format makes for very interesting reading.

531 Trost, Cathy. **Elements of Risk**. New York: Times Books, 1984, 337p.

The author investigates the history and operations of the American chemical industry by using the Dow Chemical Company as an example. This manufacturer began by making bromine and expanded by creating the pesticide DBCP and the herbicide 2,4,5-T, which contained dioxin. Both products are now banned. The author also tells the stories of a man sprayed with dioxin by a Forest Service helicopter and of a man who worked with DBCP. The human side of the accounts contrasts sharply with the corporate backdrop. This is an excellently researched book that illustrates the relationships among technology, the environment, and the individual.

532 United States Congress, Office of Technology Assessment. **Pesticide Residues in Food: Technologies for Detection**. OTA-F-398, Washington, DC: U.S. Government Printing Office, 1988, 232p.

This publication by the United States Congress is a thorough investigation and summary of the methods by which pesticide levels in food are determined. Topics include: contemporary analytical techniques; immunoassay; automation in the laboratory; and the development of new methods for pesticide detection. The book contains numerous figures and references. A glossary and list of abbreviations and acronyms are particularly useful. This is recommended to individuals concerned with the accuracy and precision of the measurements by which safe levels of pesticide residues are defined.

533 Van Den Bosch, Robert. **The Pesticide Conspiracy**. New York: Doubleday, 1978, 226p.

The author, a professor of entomology at the University of California, has written an alarming book about the pesticide industry and the people who control it. He argues that the pollution of the environment caused by these chemicals could have been avoided by the use of ecologically sound alternatives. The men who manage the industry, however, have not allowed that to happen by targeting the U.S Department of Agriculture, the Environmental Protection Agency, universities, and the media. The book, written in the late 1970s, refers to practices common two decades ago. An updated version of the story would be interesting to see if the increased environmental awareness of the American consumer has had any effect on "business as usual." Notes and a select bibliography are included.

534 Van Strum, Carol. **A Bitter Fog: Herbicides and Human Rights**. San Francisco: Sierra Club Books, 1983, 320p.

This book tells the stories of ordinary people fighting the use of herbicides in their neighborhoods. The author began her investigation of herbicides after her children were accidently sprayed by a weed-control truck. Later she noticed increased incidences of health problems, including miscarriages, birth defects, and cancers, among the members of her community. The appendices are particularly informative, covering topics such as symptoms following herbicide exposure, a warning list of pesticides, and a list of resource organizations. The personal involvement of the author makes this an interesting and emotional account.

535 Verrett, Jacqueline, and Jean Carper. **Eating May Be Hazardous to Your Health: How Your Government Fails to Protect You from the Dangers in Your Food**. New York: Simon & Schuster, 1974, 256p.

The senior author, who has a Ph.D. in biochemistry, was a researcher for the Food and Drug Administration (FDA) for over 15 years before she collaborated on this book, lending it credibility as an expose of corruption and waste within the FDA. The authors detail the problems in the testing procedures and list the dangers of specific additives. The foods in which the additives are found also are described. Suggestions for protecting one's health conclude the book. The appendix contains the filth guidelines for food as of January 1973, which most people will find eye-opening. Although outdated, this book is still an excellent reference.

536 Weir, David. **The Bhopal Syndrome**. San Francisco: Sierra Club Books, 1987, 224p.

Seventy-five percent of the U.S. population lives near a chemical plant. This volume describes the spread of the chemical industry and illustrates the increased threat the use of pesticides is to the environment. Citing the tragedy in Bhopal, India, as an example of things to come, the author examines the poor safety records of chemical plants where explosions and leaks of toxic materials are common. The pesticide content of food also is discussed. The author concludes by presenting a variety of solutions. This is a well-written overview of the subject that everyone will read.

537 Weir, David, and Mark Shapiro. **Circle of Poison: Pesticides and People in a Hungry World**. San Francisco: Institute for Food and Development Policy, 1981, 99p.

This is an intriguing account of the international sales of pesticides and their subsequent return to the United States on imported food. Some of the pesticides are among the most hazardous. In addition to telling the story of the international trade in pesticides, the authors describe the various chemicals and their known health effects. One cannot help but be a bit cynical when it is clear that economics and politics rule the day. This is a well-written and thoroughly researched book.

538 Wilcox, Fred. **Waiting for an Army to Die: The Tragedy of Agent Orange**. New York: Vintage Books, 1983, 240p. Reprinted in 1989 by Seven Locks Press with a new introduction by the author.

This is the story of Vietnam veterans who were exposed to Agent Orange for weeks at a time and who are now sick and dying. Birth defects are common in their children. The author also recounts the struggles the Vietnam veterans have had with the U.S. government over health care and disability payments for exposure to Agent Orange. The appendices cover a variety of topics including the history of 2,4,5-T and a list of organizations and individuals who will provide information on Agent Orange. This is a moving account of a tragic story.

539 Williams, Carroll M. "Third-Generation Pesticides," in *Scientific American*, **Man and the Ecosphere**. San Francisco: W. H. Freeman, 1971, p. 247-251.

This article first appeared in *Scientific American* in July 1967. The author discusses the possibility of creating pesticides that target specific insects; human and other animal populations would not be affected by these substances. In addition, the target insects cannot easily develop resistance to these third-generation pesticides. The substance ensures that the target insects' natural enemies would not be exterminated, a problem the farmer encounters when he uses general pesticides such as dieldrin or Aldrin. The article includes a brief history of pest control and cites the example of DDT.

540 Wines, Richard A. **Fertilizer in America: From Waste Recycling to Resource Exploitation**. Philadelphia: Temple University Press, 1987, 280p.

This is a history of fertilizer use in America from the early 1800s to the present. In 1845, *guano* was collected from islands 10 miles off the coast of Lima, Peru, for sale in the U.S. and Europe. Within 30 years, 10 million tons had been taken. A map from the 1880 census showed commercial fertilizer use was heaviest near Norfolk and Boston. Good discussions of the fertilizer manufacturing process are included.

12

Radiation

The world was alerted suddenly to the risks of nuclear radiation in 1979 during the near meltdown at the Three Mile Island plant in Pennsylvania and again in the spring of 1986 during the explosion of a reactor at the Chernobyl facility in the former Soviet Union. These were not the first serious nuclear incidents: Two occurred in 1957, one at Kyshtym in the U.S.S.R. and the other at Windscale in Great Britain. The Chernobyl accident, however, was the most devastating: 29 people died from radiation exposure in the first few months and hundreds suffered from radiation illness. The area became a toxic wasteland that must be quarantined for hundreds of years. As the cloud of radioactivity released from Chernobyl spread over Europe, hundreds of millions of people faced contaminated produce and milk. Tainted food appeared in the stores for months as the livestock grazed in the fallout-covered pastures. Estimates are that over 100,000 people will contract cancer as a direct result. Assurances from the U.S. Department of Energy that a similar accident could not happen in the United States did little to assuage the apprehension of the American people or to temper the anti-nuclear movement. How reliable is the technology? How safe are federal and commercial nuclear power plants? These are questions that need to be answered.

Radioactive substances are released into the environment not only during serious accidents at nuclear power plants but also during the routine manufacture of nuclear weapons and the disposal of the radioactive by-products at federal facilities. Investigations by government groups uncovered high levels of radiation in the groundwater and soil near installations at: Hanford, Washington; Cincinatti, Ohio; the Savannah River plant in Aiken, South Carolina; and the well-publicized Rocky Flats, between Denver and Boulder, Colorado. Much of the contamination arose from shallow burial pits that were used to dump low-level waste—substances that remain radioactive for tens to hundreds of years—in the 1950s and 1960s and from damaged storage tanks that leaked hundreds of thousands of gallons of radioactive substances into the groundwater. The problem is intractable; no solution exists for the long-term disposal of low-level radioactive waste. An additional threat is the disposal of high-level radioactive waste, such as uranium mill tailings and processed plutonium fuel, that remains hazardous for thousands of years.

Much of this material is in temporary storage, awaiting burial between 1,000 and 4,000 feet underground in the repository at Yucca Mountain, Nevada, a facility designed to keep 70,000 tons of waste isolated for 10,000 years.

Although the radioactivity emitted during nuclear reactions is the most dangerous, it is but one type of radiation. Others include that given off by x-rays, microwaves, and low-level alternating-current (AC) electromagnetic fields, which are associated with computers, power lines, and some electrical appliances. Overexposure to x-rays clearly has long-term consequences. The issue of the health effects of radiation from microwaves and low-level alternating-current electromagnetic fields, however, is controversial. Some scientists argue that the radiation from these sources can be linked to cataracts, miscarriage, and cancer. Other researchers find no health risks. The difference is that nuclear radiation is "ionizing"; it has enough energy to remove electrons from molecules and to break chemical bonds, whereas microwaves and low-level alternating-current electromagnetic fields are "non-ionizing"; they have only enough energy to rotate or vibrate molecules.

The works in this chapter discuss the broad range of radiation topics. Some focus on the manufacture of nuclear weapons (note Rosalie Bertell, A. Cohen and S. Lee) and atomic testing (note Howard Ball, Thomas H. Saffer and Orville E. Kelly, Philip L. Fradkin), whereas others concentrate on computer terminals and power lines (note Paul Brodeur, Elizabeth A. Scalet). Good general introductions include: "Tomorrow Always Comes" by James Bellini, *The Nuclear Waste Primer, A Handbook for Citizens* by the League of Women Voters Education Fund, *Nuclear Energy* by Karl O. Ott and Bernard I. Spinrad, *The Silent Intruder* by Charles Panati and Michael Hudson, and *The Plutonium Business and the Spread of the Bomb* by Walter C. Patterson. For those interested in an overview of the nuclear industry, Robert Deutsch's *Nuclear Power: A Rational Approach* is recommended. Other issues are: the climatic effects of nuclear explosions and war (note Lydia Dotto, Paul R. Ehrich and others, Owen Greene and others); radioactive waste disposal (note Douglas G. Brookins, Luther J. Carter, Gary E. McCuen); health consequences of the nuclear industry (note George W. Beadle and Harvey Wasserman and Norman Solomon); nuclear accidents (note Clyde W. Burleson, John May); Three Mile Island (note Daniel F. Ford, Mike Gray and Ira Rosen, Robert Leppzer); Chernobyl (note Dr. Robert Peter Gale and Thomas Hauser, David R. Marples); microwaves (note Kenneth R. Foster and Arthur W. Guy, Nicholas H. Steneck); the pros and cons of nuclear power (note Michio Kaku and Jennifer Trainer, Julie S. Bach and Lynn Hall, Samuel McCracken); steady magnetic fields (note G. Maret and others); x-rays (note Harvey Wasserman and Norman Solomon); and the Nuclear Regulatory Commission (note the Union of Concerned Scientists).

541 Bach, Julie S., and Lynn Hall (eds.). "Is Nuclear Power an Acceptable Risk?" in Bach, Julie S., and Lynn Hall (eds.), **The Environmental Crisis, Opposing Viewpoints**. St. Paul, MN: Greenhaven Press, 1986, p. 131-170.

This chapter in a book from the Opposing Viewpoints series discusses the merits and problems of nuclear power. Three topics are presented with two authors arguing the pros and cons of each subject. The topics are "America Needs More Nuclear Power Plants"; "Strategies for Dismantling Nuclear

Power Plants Are Needed"; and "Nuclear Energy Should Be Abandoned." The format makes for some lively reading and allows all opinions to be voiced.

542 Ball, Howard. **Justice Downwind: America's Atomic Testing Program in the 1950s**. New York: Oxford University Press, 1986, 280p.

The author traces the development of atomic testing in the United States from 1951 to 1963. During those years, bombs were detonated above ground by the U.S. government at the Nevada Test Site, approximately 100 miles north of Las Vegas. Bombs were exploded only when the wind blew east, away from California and Las Vegas. The subsequent ill health of the "downwinders" caused them to sue the government. The author tells the story of the "downwinders" compellingly. In doing so, he illustrates the effects of nuclear radiation and atomic fallout. Numerous figures and a good index accompany the text. The book is well researched and scholarly.

543 Beadle, George W. "Ionizing Radiation and the Citizen," in *Scientific American*, **Man and the Ecosphere**. San Francisco: W. H. Freeman, 1971, p. 155-161.

This article first appeared in *Scientific American* in September 1959. The author, a Nobel laureate, attempts to outline the effects of radiation and the philosophy of setting standards. The emphasis is on mutations of human cells. All mutations are considered harmful and any dose of radiation is believed to cause mutations. The author argues that the amounts of ionizing radiation should be reduced and that society should not wait until the answers to problems are known before making decisions concerning radiation. The figures that accompany the text are excellent.

544 Bellini, James. "Tomorrow Always Comes," in Bellini, James, **High Tech Holocaust**. San Francisco: Sierra Club Books, 1986, p. 59-96. ****O**

This chapter in a book that argues the evils of our modern technological society is a good introduction and overview to radiation from all sources. The chapter begins with a story about Canonsburg, Pennsylvania, known as Radiation City, where incidences of leukemia and other cancers were far above average. The author also describes the biological effects of radiation, the history of nuclear power, the development of x-rays, and the amount of radiation waste that existed in the mid-1980s. A table of the health effects of different doses of radiation is included.

545 Bertell, Rosalie. **No Immediate Danger: Prognosis for a Radioactive Earth**. Summertown, TN: Book Publishing, 1985, 435p.

The author argues that escalating radiation levels from the manufacture of nuclear weapons and the generation of nuclear power has led to the deaths of 16 million people. She examines why governments and scientists state that nuclear fuel is safe and that the public should not be afraid. A good introduction to nuclear radiation and its effects begins the volume. Nuclear policy in the military and civilian sectors also are discussed. Extensive references and a good index are included. This is an alarming account of an everyday problem.

546 Brodeur, Paul. **Currents of Death: Power Lines, Computer Terminals, and the Attempt to Cover up Their Threat to Your Health**. New York: Simon & Schuster, 1989, 333p. ****S**

This thoroughly researched and scholarly article about low-level alternating-current electromagnetic fields should open a few eyes to the potential health problems associated with computers, power lines, and electric blankets. The intrigue of the investigative style serves the book well and holds the reader's attention through many dry facts. Some horrific statistics are cited, including one out of every four homes in the United States may be near high-tension wires that radiate hazardous magnetic fields, children who live proximal to high-tension wires develop cancer twice as often as children who live away from the magnetic fields, and people who operate certain types of electronic equipment are stricken with cancer at a greater rate than the general population. The study is broken into three parts: 1) "Power Lines"; 2) "Cover-up"; and 3) "Computer Terminals." Sections of the book originally were published in *The New Yorker*. An excellent index makes specific discussions easy to find.

547 Brookins, Douglas G. (ed.). **The Geological Disposal of High-Level Radioactive Wastes**. Athens, Greece: Theophrastus Publications, 1987, 606p.

This volume contains 27 papers by 40 authors from 11 countries and presents an international overview of the disposal of high-level radioactive waste. The book contains five sections: 1) "The High-Level Radioactive Waste Isolation Program"; 2) "Waste Forms"; 3) "Leaching and Retardation"; 4) "Natural Analogues"; and 5) "Evaporites and Sub-Seabed Disposal." The text has numerous illustrations, many photographs, and complete reference lists. The volume is more intelligible to the reader who has a background knowledge of chemistry, physics, and mathematics.

548 Burleson, Clyde W. **The Day the Bomb Fell on America, True Stories of the Nuclear Age**. Englewood Cliffs, NJ: Prentice-Hall, 1978, 261p.

The author has compiled a series of true stories based on classified government documents from the United States and other countries, articles, and eyewitness accounts. He cites the accidental release of an atomic weapon (that did not detonate) on the town of Mars Bluff, South Carolina, in 1958, the death of Karen Silkwood, a worker at a nuclear power plant, on her way to report irregularities at the plant in 1974 (an incident made famous by the movie *Silkwood*), the contamination of large areas in the Ural Mountains of the former Soviet Union during two major nuclear disasters in the 1950s, and the deliberate release of radioactive clouds into the air by officials at the Windscale plutonium plant in England in 1957. The stories are chilling. The book does not have a table of contents or index, which makes finding specific information difficult.

549 Carter, Luther J. **Nuclear Imperatives and Public Trust, Dealing with Radioactive Waste**. Washington, DC: Resources for the Future, 1987, 473p.

The author, a science journalist, argues that putting radioactive nuclear waste into repositories deep in the ground is the solution to prevention of illegal use of plutonium and to protection of human health. Public opposition and "technological uncertainty" have ensured that these repositories will not be built. The book is divided into four parts: 1) "Sources of Public Unease"; 2) "Searching for a Waste Policy"; 3) Europe, Japan and the International Waste Problem"; and 4) "A Time to Act." A glossary, name index, and subject index conclude the book. This is a scholarly and objective view of an important subject. The text is thoroughly footnoted.

550 Chin-Fu-Tsang (ed.). **Coupled Processes Associated with Nuclear Waste Repositories**. Orlando, FL: Academic Press, 1987.

This volume covers an international symposium on nuclear waste repositories and uses concepts from geology, hydrology, chemistry, and geophysics in its evaluation of the sites. The text discusses field studies of geohydrologic, geomechanical, and geochemical processes. Also included is an informative discussion by the agencies involved in nuclear waste disposal. The focus is on siting problems for waste repository facilities. The approach is very technical but the introduction is useful to the general reader.

551 Cohen, A., and S. Lee (eds.). **Nuclear Weapons and the Future of Humanity: The Fundamental Questions**. Totowa, NJ: Rowman and Allanheld, 1986, 496p.

This is a collection of chapters written by experts in the field of nuclear weapons and the effects of nuclear radiation. One of the most interesting is by the physicist, John P. Holdren, entitled "The Dynamics of the Nuclear Arms Race: History, Status, Prospects." The warnings are horrifying. This book should be read by everyone.

552 Comar, Cyril L. "Biological Aspects of Nuclear Weapons," in Bresler, Jack B. (ed.), **Human Ecology, Collected Readings**. Reading, MA: Addison-Wesley, 1966, p. 363-377.

The material in this chapter was given initially as a lecture on nuclear peril and disarmament. The author paints a balanced picture of the biological problems of radiation but oversimplifies and omits details. Problems derive from fallout and contamination and include either genetic or somatic effects. "Threshold" levels of exposure are discussed. Low and high levels of radioactive exposure are evaluated. Contributions of medical procedures also are estimated and are assumed equivalent to the background levels of radiation on the Earth. Some solutions for highly contaminated areas are presented.

553 Deutsch, Robert W. **Nuclear Power: A Rational Approach, 4th Edition**. Columbia, MD: GP Courseware, 1987, 93p. ****O**

The author, the director of a company that provides services to the nuclear industry, has attempted to present an objective account of nuclear power. Topics include: a history of nuclear power; regulation of the industry; the Three Mile Island accident; Chernobyl; risk analysis; radioactivity; and nuclear fuel. The chapters of greatest interest are "How Dangerous Is Radiation?"; "Disposal of Radioactive Waste"; and "Radiation Sources and Management." Numerous figures accompany the text. This is a good overview of the nuclear industry for the nonexpert.

554 Dotto, Lydia. **Planet Earth in Jeopardy**. Chichester, England: John Wiley, 1986, 134p.

This book is a nontechnical summary of a comprehensive scientific report prepared by the Scientific Committee on Problems in the Environment (SCOPE) on the long-term environmental consequences of nuclear war. Some conclusions are that starvation is a serious risk for the majority of people on Earth after a large-scale nuclear war, climate changes will affect the entire Earth for several years, and the global environment and social systems may fragment such that "recovery to pre-war conditions" is impossible. A reference list is included. This is a good source of information for the individual interested in the scientific debate behind nuclear winter.

555 Ehrlich, Paul R., Carl Sagan, Donald Kennedy, and Walter Orr Roberts. **The Cold and the Dark**. New York: W. W. Norton, 1984, 229p.

The book, the result of a year-long project involving 200 scientists from many countries, describes the effects of nuclear war on the environment. "Nuclear winter," whereby the planet is shrouded in smoke and dust generated by the detonation of nuclear weapons and the subsequent firestorms to such an extent that sunlight is blocked, would lower global temperatures significantly. Agriculture could not proceed. Famine would be widespread. This discussion of the climatic and biological consequences of nuclear war is essential reading. Of particular interest is the chapter entitled "The Moscow Link: A Dialogue Between U.S. and Soviet Scientists."

556 Epstein, William. "A Ban on the Production of Fissionable Material for Weapons," in *Scientific American*, v. 243, n. 1, July 1980, p. 43-51.

This discussion of arms limitations agreements, including SALT II, highlights their two primary benefits: the cessation of further buildup of nuclear arsenals and the prevention of the transfer of nuclear weapons technology to non-nuclear countries. The article is full of statistics on world weapons inventories. Concise descriptions of nuclear materials are given in figure captions. This is a well-illustrated and informative source for the general reader.

557 Fetter, Steven A., and Kosta Tsipis. "Catastrophic Releases of Radioactivity, " in *Scientific American*, v. 244, n. 4, April 1981, p. 41-47.

The authors describe the amounts of radioactivity that are released during three different events: 1) limited nuclear war; 2) inadvertent release of gas from a nuclear power plant; and 3) an accident while manufacturing or transporting a nuclear weapon. Maps of "dose contours" illustrate the trajectories of the radiation and the amounts of fallout in the three scenarios. The conclusion is that a single nuclear weapon would have a far more devastating effect than the worst possible accident at a nuclear power plant.

558 Ford, Daniel F. **Three Mile Island: Thirty Minutes to Meltdown**. New York: Penguin Books, 1981, 271p.

The author, a former executive director of the Union of Concerned Scientists, specializes in nuclear policy. In his book, he follows the history of the nuclear power plant at Three Mile Island and explains why the accident happened. The reasons include cost-cutting procedures allowed by federal regulations, lack of preparedness of the plant operators, and mismanagement by the Nuclear Regulatory Commission. The recounting of the Three Mile Island accident as it occurred is engrossing. This is an interesting book that uses government documents and the records of plant designers as its sources.

559 Foster, Kenneth R., and Arthur W. Guy. "The Microwave Problem," in *Scientific American*, v. 255, n. 3, September 1986, p. 32-39.

The authors discuss the effects of exposure to low levels of microwaves and the limits that should be established. Their main point is that the data are ambiguous. The history of the issue is traced. An introduction to microwaves is included.

560 Fradkin, Philip L. **Fallout: An American Nuclear Tragedy**. Tucson, AZ: University of Arizona Press, 1989, 300p.

The author reviews the trial of *Irene Allen* v. *the United States* in which the plaintiff attempted to force the government to accept responsibility for the high incidence of cancer in southern Nevada where above-ground testing had occurred in the 1950s and 1960s. The case focused national attention on the problem of fallout. The book also gives a good overview of the development of nuclear weapons and their subsequent atmospheric testing. Slot Harry, the most radioactive of the above-ground tests, receives the greatest coverage. This is an informative treatment of nuclear testing although the trial descriptions can get too involved.

561 Freeman, Leslie J. **Nuclear Witnesses, Insiders Speak Out**. New York: W. W. Norton, 1981, 330p.

This is a collection of first-person accounts about the effects of radiation on people's lives told by those who have worked in the nuclear power industry. The stories are gripping, emotional, and shocking. Each chapter takes the reader to a different setting: the plutonium factory at Hanford, Washington; the uranium mines on a Navajo reservation (a particularly wrenching testimony revealed by three widows); the weapons facility at Lawrence-Livermore Laboratory in California; and the Indian Point, Dresden, and Pilgrim nuclear power plants. An afterword by Dr. Helen Caldicott, an anti-nuclear activist, a bibliography, and an index also are included.

562 Gale, Dr. Robert Peter, and Thomas Hauser. **Final Warning**. New York: Warner Books, 1988, 230p.

This is an engrossing eyewitness account of a doctor's trip to the former Soviet Union immediately after the Chernobyl nuclear accident. The doctor met with President Mikhail Gorbachev and visited the Soviet hospitals where the injured were being treated. The book includes discussions of the need for international cooperation and the future of nuclear power. Descriptions of the aftermath of Chernobyl, both in its human toll and its destruction of the landscape, are given. The author has written an overview of the Chernobyl incident that is both horrifying and uplifting.

563 Gough, William C., and Bernard J. Eastlund. "The Prospects of Fusion Power," in *Scientific American*, **Man and the Ecosphere**. San Francisco: W. H. Freeman, 1971, p. 252-266.

This article first appeared in *Scientific American* in February 1971. The authors give an excellent introduction to fusion power (the energy of the stars). They begin by outlining available energy options and describing the importance of energy in the industrialized world. The history of fusion studies is detailed. Additional topics include the design of fusion reactors, environmental considerations of fusion power, and the realistic estimates of when fusion power will be feasible. The figures are very informative. One illustration depicts 10 fusion reactions that are potentially useful in fusion reactors.

564 Gould, Jay M., and Benjamin A. Goldman. **Deadly Deceit, Low Level Radiation, High Level Cover-Up**. New York: Four Walls Eight Windows, 1990, 222p. **F

The authors look at the statistics, based on government and industry records, of the relationship between atomic radiation levels and mortality rates. The book has three main sections: case studies, which include Chernobyl, Savannah River, and Three Mile Island; implications, which encompass atmospheric bomb tests, infant mortality and milk, and radiation and aids; and conclusions. Numerous figures complement the text. The authors present considerable evidence to support their alarming findings. This is a fascinating and alarming book that should be read by everyone.

565 Gray, Mike, and Ira Rosen. **The Warning**. New York: W. W. Norton, 1982, 287p.

The authors, one the author of *China Syndrome* and the other a producer for "60 Minutes," have written an account of the Three Mile Island nuclear accident based on interviews with the people who operated the power plant. They conclude that mechanical failure in conjunction with human error led to the event. The tragedy is that the problem had been identified months before the accident. Because the book unfolds like a novel, the story is suspenseful and engrossing. The short glossary is extremely valuable.

566 Greene, Owen, Ian Percival, and Irene Ridge. **Nuclear Winter**. Cambridge, England: Polity Press, 1985, 216p.

The authors discuss the possible effects of a large-scale nuclear war on the global environment. The emphasis is on fires, smoke, and particulates in the atmosphere that would lead to blocking of the sun and cooling of the climate, hence, nuclear winter. Throughout the book, the authors stress that although much research suggests nuclear winter would occur, it is not scientifically proven. Their hope is that the proof will never appear. This is an excellent source for the general reader. Numerous illustrations accompany the text.

567 Hatheway, Allen W., and Cole R. McClure, Jr. **Geology in the Siting of Nuclear Power Plants**. Reviews in Engineering Geology IV. Boulder, CO: Geological Society of America, 1979, 263p.

This book covers a wide variety of topics, including the regulatory process and licensing, site evaluation, the state-federal partnership, seismicity of the United States, and errors in determination of earthquake locations. Techniques used to evaluate sites also are discussed. These are groundwater studies, remote-sensing, and borehole geophysics. In addition, some of the common pitfalls in site evaluation are described. The text is informative but is very much biased in favor of the nuclear industry.

568 Hilgartner, Stephen, Richard C. Bell, and Rory O'Connor. **Nukespeak: The Selling of Nuclear Technology in America**. New York: Penguin Books, 1982, 282p.

The authors, citing George Orwell as their inspiration, attempt to expose the language used by nuclear developers to illustrate the "sunny side of the bomb." Examples are phrases such as "a peaceful nuclear device," "a potential nuclear explosive" (plutonium), and an "event" (nuclear accident). The book has eight parts: 1) "Nuclear Visions"; 2) "The Ministry of Truth"; 3) "Radiation: Managing the Political Fallout"; 4) "One Chance in Five Billion"; 5) "Leaks, Spills, and Promises"; 6) "Material Unaccounted For"; 7) "There Isn't Any Choice"; and 8) "Atoms for War." An index of nukespeak words is included. The text has numerous illustrations. This is an interesting and enlightening book that helps to demystify the terminology used in the nuclear industry.

569 Kaku, Michio, and Jennifer Trainer (eds.). **Nuclear Power: Both Sides, the Best Arguments for and Against the Most Controversial Technology**. New York: W. W. Norton, 1982, 279p.

Six of the seven chapters in this book contain papers that debate a central theme. Topics are: radiation; reactor safety; nuclear waste disposal; economics; alternatives to nuclear fission; and the future of the nuclear industry. The other chapter is a brief overview of the history of nuclear technology development. The authors' purpose was to allow citizens to hear all perspectives and make an informed decision.

570 Koval, Thomas M. (ed.). **Environmental Radioactivity**. Bethesda, MD: National Council on Radiation Protection and Measurements, 1984, Science of the Total Environment Series, v. 35, no. 3, p. 207-445.

This collection of papers from the annual meeting of the National Council on Radiation Protection and Measurements addresses topics as diverse as radon in the home and contaminated atolls of the Marshall Islands. Other subjects are: Three Mile Island; fallout on the Nevada Test Site; and the radiological risk of waste-disposal sites. Twelve papers, a review lecture, and brief summaries of important publications compose the volume. The approach is suitable for the general reader.

571 The League of Women Voters Education Fund. **The Nuclear Waste Primer: A Handbook for Citizens**. New York: Nick Lyons Books, 1985, 90p. ****O**

This practical guide contains much of the information needed to understand the nuclear waste problem in the United States. Topics are the sources of radioactive waste, types of radiation hazards and their biological effects, waste management, the politics of waste management, possible solutions, and the role of the citizen. A list of organizations involved in the disposal and regulation of nuclear waste is included. The short glossary is excellent. The text is straightforward.

572 Leppzer, Robert. **Voices from Three Mile Island: The People Speak Out**. Trumansburg, NY: Crossing Press, 1980, 86p.

This intriguing book is a collection of conversations among people who lived near Three Mile Island during the accident at the nuclear power plant on March 28, 1979. The author asks questions occasionally to clarify points. The book begins with a description of the people. Conversations are grouped into chapters with titles such as "The Accident," "Suspicion," "Frustration," and "Never Again." This intensely personal account was broadcast initially as a two-hour radio documentary, "Early Warnings: Voices from Three Mile Island."

573 Lewis, Harold W. "The Safety of Fission Reactors," in *Scientific American*, v. 242, n. 3, March 1980, p. 53-65.

This lucid and informative article discusses the safety of nuclear fission power plants to allay fears induced by the Three Mile Island accident. The author states the plants can be considered "perfectly safe" because "no member of the public has ever been injured by a commercial reactor accident" (this was before the Chernobyl disaster in the former Soviet Union). The workings and safety features of a reactor are described and illustrated.

574 Lipschutz, Ronnie D. **Radioactive Waste: Politics, Technology, and Risk**. Cambridge, MA: Ballinger, 1980, 246p.

The author is a physicist who later joined the Union of Concerned Scientists and edited their newsletter. He has written a lucid, nontechnical treatment of the problems surrounding the management and disposal of radioactive waste. The book has six chapters. The first two provide an introduction to radioactivity and the nuclear fuel cycle. The next three discuss the management, storage, and disposal of radioactive wastes in the past and the present. The last chapter offers suggestions for a successful program in waste management. Three appendixes, an extensive glossary, a list of references, a bibliography, and an index also are included.

575 Lovins, Amory B., and John H. Price. **Non-nuclear Futures**. San Francisco: Friends of the Earth International and Cambridge, MA: Ballinger, 1975, 223p.

The authors consider an ethical energy strategy for the future and conclude that ridiculous amounts of capital are required to ensure that nuclear power is safe. The book does not provide an overview of nuclear engineering but does give readers knowledge of the basic technical, social, economic, and policy issues that allows them to make a decision about nuclear power. The text is well illustrated and includes several appendices that contain detailed information on specific aspects of the nuclear industry and the nuclear power plant. Chapter 5 on waste management and chapter 6 on plutonium toxicity are particularly good. The authors assume no previous experience in mathematics or science.

576 Maret, G., J. Kiepenheuer, and N. Boccara (eds.). **Biophysical Effects of Steady Magnetic Fields: Proceedings of the Workshop, Les Houches, France, February 26-March 5, 1986**. Berlin: Springer-Verlag, 1986, 231p.

This is a collection of technical scientific papers from a meeting in France in 1986 that brought together physicists, chemists, doctors, physiologists, and biologists to discuss the health effects of steady-state magnetic fields. The papers describe the basic response of living matter on a molecular level. Several papers focus on the effect of the Earth's magnetic field on migratory animals. Others concentrate on Nuclear Magnetic Resonance Imaging, now called Magnetic Resonance Imaging (MRI), a technique used in today's hospitals.

577 Marples, David R. **The Social Impact of the Chernobyl Disaster**. London: Macmillan, 1988, 313p.

The author looks at the aftermath of the Chernobyl nuclear accident in human terms. He traveled to the former Soviet Union in 1987 to discover the extent of the contamination, the number of people who had died, the future of the Soviet nuclear industry, and the reactions of people who were there. The introduction, written by Dr. Victor G. Snell, the manager of safety engineering at a project of Atomic Energy of Canada, Limited, summarizes the cause of the accident. Chapter titles are "The Victims of Chernobyl"; "The Environmental Impact"; "The Economic and Political Repercussions"; "Images of Chernobyl: Arts and the Public"; "The Special Zone"; "Restoration and Reconstruction"; and "The Nuclear Power Debate." Extensive notes, a list of Soviet and non-Soviet references, and an index accompany the text.

578 Martin, Daniel. **Three Mile Island, Prologue or Epilogue?** Cambridge, MA: Ballinger, 1980, 251p.

The author, a former professor of public administration at the University of Baltimore, summarizes the opening hours of the accident at the Three Mile Island nuclear power plant on March 28, 1979, based on the testimonies that plant workers, employees of the Pennsylvania emergency system, and federal observers gave before the United States Congress. Included in his description are the responses of the state and federal governmental agencies. An explanation of the way a reactor works allows the reader to understand the nature of the mechanical failure that led to the accident. The future of the nuclear industry also is discussed. A glossary, list of acronyms, and an index accompany the text.

579 May, John. **The Greenpeace Book of the Nuclear Age: The Hidden History, the Human Cost**. New York: Pantheon Books, 1989, 378p. ****F**

This book claims it is the "most comprehensive record of nuclear accidents and radiation incidents ever published" and appears to live up to its billing. The list of events begins in the 1940s and continues to the 1980s. The first part of the book details the risks associated with nuclear power and explains the fundamentals of nuclear physics. The text is well researched. A thorough index concludes the book. This is recommended to anyone who needs to know specific facts about any nuclear incident.

580 McCracken, Samuel. **The War Against the Atom**. New York: Bantam Books, 1982, 206p.

This is a book that states the case for nuclear power. The author, a former senior research fellow at the Center for Energy at Boston University, argues that nuclear power poses fewer environmental and health risks than alternative energy sources such as coal. The Three Mile Island accident is analyzed and viewed as less serious than originally believed. To justify his case for nuclear power, the author also tries to understand why the anti-nuclear movement is so strong and vocal. The bibliography has some references, which are described briefly. This is a controversial book that deserves to be read.

581 McCuen, Gary E. **Nuclear Waste: The Biggest Clean-Up in History**. Hudson, WI: Gary E. McCuen Publications, 1990, 150p.

Part of the Ideas in Conflict series, this volume contains 19 essays by a variety of authors on all aspects of the problem of nuclear waste disposal. The first chapter presents an overview of the problem. The remaining four chapters discuss the issues of military radioactive waste, waste storage, decommissioning nuclear power plants, and the nuclear weapons complex. At the end of each chapter are activities designed to improve the reader's reasoning skills. Two bibliographies are included.

582 McVay, G. L. **Scientific Basis for Nuclear Waste Management VII**. Materials Research Society Volume 26. New York: Elsevier Science, 1985, 1121p.

This volume contains 122 papers presented at a symposium in 1983 that addressed the disposal of high-level nuclear waste. The researchers focus on groundwater flow and radionuclide transport. The book is very technical, consisting of work done by geologists and materials engineers. The reader should have a good quantitative background and familiarity with chemical and geological principles. The reference list and the tables of data, however, may be of interest to the general reader.

583 Meehan, Richard L. **The Atom and the Fault: Experts, Earthquakes, and Nuclear Power**. Cambridge, MA: MIT Press, 1984, 161p.

The author uses the controversy around the first corporately owned nuclear test reactor in the United States near Vallecitos, California, to illustrate the problems that arise whenever new technologies are implemented. In this case, "experts" were divided on the issue of the earthquake danger near the power plant. Some saw abundant evidence for an earthquake fault, whereas others observed only an ancient landslide. What is to be done when decisions affect the safety of tens of thousands of people? The book is well written and scholarly. A short, annotated bibliography is included.

584 Miller, Richard L. **Under the Cloud: The Decades of Nuclear Testing**. New York: Free Press, 1986, 547p.

This book traces the history of the above-ground nuclear testing of the 1950s and 1960s in the desert southwest of the United States, beginning with "Trinity," the first atomic bomb explosion in 1945. The authors tells stories of school children sent up on hillsides to watch the flash of light given off by the bombs as they detonate; of troops sent into ground zero immediately after the explosion; of "Doom Towns" built and populated by mannequins; of pigs dressed in army uniforms and placed near the target to test fiber durability; and of public relations blitzes designed to awe the American public. The most frightening, however, are the paths of the radioactive clouds over the country, illustrated in a series of maps. Radiation levels in cities across the country after nuclear tests and radiation track descriptions are listed in two appendices. This is an alarming book that deserves to be read.

585 Milnes, A. G. **Geology and Radwaste**. London: Academic Press, 1985, 328p.

This volume is divided into three parts. Part One gives background information on radioactive waste and disposal techniques. Part Two describes the important geological factors used to assess the disposal site and Part Three discusses the interplay between geologic insight and sociopolitical considerations in choosing a waste repository. Part One is beneficial to most readers although it is quite quantitative in its approach to waste disposal. Part Two requires some familiarity with geological concepts. Part Three also is of general interest. The book is a good introduction to the problem of waste disposal—a problem, as the author states, that needs to be solved whether or not one approves of nuclear power.

586 Murphy, Dervla. **Nuclear Stakes, Race to the Finish**. New Haven, CT: Ticknor & Fields, 1982, 264p.

The author was visiting friends in Pennsylvania when the Three Mile Island nuclear accident occurred in 1979. Realizing the dangers inherent in the technology, she investigated the nuclear power and weapons industry. The book begins with the history of the first atomic bomb. Other topics are the people involved with the nuclear movement (both pro and con), "safe" radiation levels, and proliferation. The tone of the book is urgent: the author states at the outset that nuclear weapons and power have the potential to "exterminate all life on Earth." The glossary and bibliography are useful.

587 Murray, Raymond L. **Understanding Radioactive Waste**. Columbus, OH: Battelle, 1985, 128p. ****F**

This unbiased resource on radioactive waste discusses all aspects of the problem, including waste generation; waste disposal; and safety standards. The volume contains many tables of data and statistics. The text is not technical. The work is recommended to the individual seeking factual information to support general statements and to make knowledgeable decisions.

588 National Research Council. **Radioactivity in the Marine Environment**. Washington, DC: U.S. Government Printing Office, 1971, 272p.

This is an account of what had been learned about the behavior and effect of radioactive material in the marine environment between 1957 and 1971. Topics include: sources of radioactivity such as natural nuclear explosives, nuclear reactors, and radioactive waste; distribution of radionuclides from nuclear explosions; marine chemistry; ocean currents and ocean mixing; and the accumulation and effects of radiation on marine organisms and man. The introduction summarizes the contents and gives a brief historical account of radioactivity in the marine environment.

589 National Research Council, Committee on Vision. **Video Displays, Work and Vision**. Washington, DC: National Academy Press, 1983, 273p.

This report is the result of a National Research Council review of measurements of emissions from video display terminals (VDTs). The biological effects of radiation are detailed and radiation hazards of VDTs are debated. The study also includes discussions of the relationship among cataracts, stress, and various muscular ailments and VDTs.

590 **Nuclear Power in an Age of Uncertainty**. Washington, DC: U.S. Congress, Office of Technology Assessment, OTA-E-216, February 1984, 293p.

This document is the result of a request by the U.S. Congress to determine the future of nuclear power in the United States and to reduce the risks associated with the technology. Topics include: alternative reactor systems; the regulation of nuclear power; survival of the nuclear industry; public attitudes toward nuclear power; and policy options. Numerous figures, graphs, and tables accompany the text. The appendix, containing acronyms and a glossary, is useful.

591 Ott, Karl O., and Bernard I. Spinrad. **Nuclear Energy**. New York: Plenum Press, 1985, 386p. ****O**

This is a collection of essays written by scientists who work primarily in universities and research institutes. The purpose is to educate the public about nuclear energy using information that is not evident in the rhetoric of the anti-technology and anti-nuclear movements. Topics include: energy and society; the economics of nuclear power; recycling and proliferation; risk assessment; and special nuclear issues. The first chapter provides an excellent overview and introduction to nuclear power. This is an interesting and scholarly work. Some essays contain diagrams and illustrations.

592 Panati, Charles, and Michael Hudson. "The Blanket of Electrical Smog," in Panati, Charles, and Michael Hudson, **The Silent Intruder**. Boston: Houghton Mifflin, 1981, 208p.

This is a chapter in a book about the effects of radiation and how to survive them. The authors discuss electrical pollution which they define as "nonionizing radiation emitted by electrical devices." Objects that emit the radiation include microwave ovens, radio and television transmitters, and high-voltage power lines. The type of electrical radiation believed to be most harmful are the Extremely Low Frequency (ELF) waves of power lines because these waves have frequencies in the same range as brainwaves. The chapter details several examples of possible injury from electrical pollution. The authors are fair and point out conflicting and absent data where necessary.

593 Panati, Charles, and Michael Hudson. **The Silent Intruder**. Boston: Houghton Mifflin, 1981, 208p. ****O**

The effects of radiation and how to survive them are addressed in this book. Most sources of radiation are discussed: ultraviolet rays from the sun; x-rays used in medical analysis; microwaves; and high-voltage power lines. Nuclear power plants also are described. Practical suggestions are given to avoid exposure. These include: limiting time in the sun; reducing the number of dental and medical x-rays an individual has taken; and not living close to electric power lines. The appendices contain useful information such as cancer risks from x-rays; radiation doses to a fetus from diagnostic x-ray tests; and average doses for diagnostic x-ray examinations.

594 Patterson, Walter C. **The Plutonium Business and the Spread of the Bomb**. San Francisco: Sierra Club Books, 1984, 288p. ****O**

The author is a consultant to Friends of the Earth, a conservation organization, and, consequently, has written an informative book accessible to the layperson. He discusses the efforts undertaken around the world to develop fast-breeder reactors that would supply almost limitless amounts of electricity through nuclear power. The danger of plutonium is described. It is not only a health hazard but a military threat. Diversion of plutonium earmarked for nuclear power plants to individuals interested in the manufacture of nuclear weapons is not only possible but probable. The appendices contain remarks made by Presidents Eisenhower, Ford, and Carter on nuclear power and policy. The bibliography is thorough.

595 Perrow, Charles. **Normal Accidents, Living with High-Risk Technologies**. New York: Basic Books, 1984, 386p.

The author, a prominent sociologist, has written a book describing the complex processes that govern modern technology and the consequences that arise when accidents occur. The first part of the work emphasizes nuclear power, using the Three Mile Island catastrophe as an example. Other topics include: nuclear weapons; the aerospace and petrochemical industries; recombinant DNA research and application; and marine technology. The author also debates the validity of the new "science" of risk assessment. Extensive notes, a reference list, and an index accompany the text. This is an interesting and thought-provoking book.

596 Roberts, L. E. J. **Nuclear Power and Public Responsibility**. Cambridge, England: Cambridge University Press, 1984, 120p.

The author is the director of one of Great Britain's nuclear research facilities. As such, he has written a book that argues in favor of nuclear power and states the opinion that the industry is the most safety-conscious of all energy industries in Great Britain. The author weighs the risks against the benefits to determine the acceptable level of safety. The book is well written, clear, and concise. The discussion of risks is illuminating. Numerous figures, a good index, and a short bibliography accompany the text.

597 Robinson, Marilynne. **Mother Country: Britain, the Welfare State, and Nuclear Power**. New York: Farrar, Straus & Giroux, 1989, 261p.

The world's largest source of radioactive contamination is Great Britain. This is directly the result of the nuclear waste reprocessing facility called Sellafield in England's Lake District. At Sellafield, nuclear waste from Japan and Western Europe is treated to extract as much plutonium and uranium as possible. The rest is dumped into the Irish Sea. A consequence is high levels of leukemia among children living near the English and Irish coasts. The author addresses this issue and condemns Britain for exploiting its people for profit. The book is provocative and interesting. The bibliography consists primarily of newspaper articles.

598 Robinson, Mark Aaron. **100 Grams of Uranium Equal 290 Tons of Coal**. Kelso, WA: R&D Engineering, 1987, 124p.

The author has analyzed "the impacts and consequences of producing electricity from both coal and nuclear fuel." The categories in which the two energy sources are compared include availability of fuel, processing, mining, waste, pollutants, safety, and economics. An advocate of nuclear technology, his conclusion is that the ill effects from coal are far greater than those from nuclear power. This book has an interesting perspective and intriguing arguments.

599 Saffer, Thomas H., and Orville E. Kelly. **Countdown Zero**. Middlesex, England: Penguin Books, 1982, 351p.

This account of above-ground nuclear testing and its health effects by two veterans who participated is gripping and horrifying. One of the authors, Thomas Saffer, died before the book was finished. A casualty of the "war games" played by the United States during the 1950s and 1960s, he witnessed 22 bomb blasts at Eniwetok Atoll in the South Pacific in 1957. The lack of precautions taken by members of the U.S. armed forces and the cover-up by the federal government are emphasized. The most sobering chapter is "The Reunion," which describes the National Association of Atomic Veterans, founded in April 1979. A list of numbers to call and organizations to contact for information for veterans, a glossary, notes, and an index are included.

600 Scalet, Elizabeth A. "VDTs and Radiation," in Scalet, Elizabeth A., **VDT Health and Safety, Issues and Solutions**. Lawrence, KS: Ergosyst, 1987, p. 7-26.

This chapter in a book on the safety of video display terminals (VDTs) is an excellent introduction to the debate about the types and effects of radiation emitted by VDTs. The discussion begins with an explanation of radiation in general and progresses to a description of specific types of radiation. VDTs are characterized by the emission of very low frequency radiation. Concern over the safety of VDTs has been raised because some studies of mice suggest a link between very low frequency radiation and fetal malformation. The last page of the chapter lists five additional sources for information on VDTs and radiation.

601 Siegel, Mark A., Carol D. Foster, and Nancy R. Jacobs. "Nuclear Waste," in **Energy: An Issue of the 90's**. Wylie, TX: Information Plus, 1989, p. 78-84.

This is a concise, informative treatment of nuclear waste. The basic types of radioactive waste are defined: uranium mill tailings; spent fuel; high-level waste; and transuranic wastes. The Nuclear Waste Policy Act of 1982 is summarized. Additional topics include weapon plants contamination and public health hazards. Specific problems also are outlined. Of particular interest is a map of the United States that shows the locations of 15 facilities that generate nuclear waste. The caption to the figure briefly describes the output of each facility.

602 Skjei, Eric, and M. Donald Whorton. "Radiation," in Skjei, Eric, and M. Donald Whorton, **Of Mice and Molecules**. New York: Dial Press, 1983, p. 233-263.

This chapter, in a book on technology and human survival, gives a good overview of radiation, its causes, and its effects. The discussion begins with an explanation of radiation: how it works and how it is measured. Effects are described next. These include: ulcers; abdominal pain; nausea; hemorrhage; cancer; and birth defects. The chapter concludes with a list of radwastes and sources of radiation exposure such as cosmic rays, medical x-rays, and nuclear power plants when gases escape or meltdown occurs. The treatment is scholarly and informative but is not technical.

603 Slesin, Louis. "People Are Antennas, Too: The Biology of the Electromagnetic Spectrum," in *Whole Earth Review*, v. 50, n. 50, Spring 1986, p. 50-55.

This article is a good introduction to the biological effects of nonionizing radiation, or radiation that does not have enough energy to remove electrons from molecules or break chemical bonds. The radiation, however, can make molecules vibrate or rotate, thereby releasing heat. Ultraviolet radiation, visible light, infrared radiation, radio waves, and microwaves are examples of nonionizing radiation. Thc author summarizes the debate between the importance of thermal (or heat) and nonthermal effects. Some scientists argue if no thermal effect exists, the nonionizing radiation is not harmful.

604 Steneck, Nicholas H. **The Microwave Debate**. Cambridge, MA: MIT Press, 1984, 279p.

The author, a professor of history at the University of Michigan, presents a scholarly and objective analysis of the health effects of microwaves. The book is divided into two parts: the private years, 1930-1967, and the public years, 1967 to the present. During the private years, the scientific community concentrated on the thermal effects of the radiation, which were believed beneficial. Athermal effects were of interest only to the military until the late 1960s when environmental and consumer activists began to worry about the possible health hazards of microwaves. Controversies surrounding the energy emitted by televisions, the location of communication antennas, and the bombardment of the U.S. embassy in Moscow by microwave radiation are discussed. Notes and an index accompany the text.

605 Sternglass, Ernest. **Secret Fallout: Low-Level Radiation from Hiroshima to Three Mile Island**. New York: McGraw-Hill, 1981, 386p.

This book examines low-level radiation in the atmosphere that resulted from nuclear power accidents or nuclear explosions. The author, a professor of radiation physics, attempts to expose the dangers of nuclear power and the apathy of the federal government toward these hazards. He cites three life-threatening properties of nuclear power: 1) nuclear power plant accidents; 2) the creation, as a by-product, of plutonium 239, one of the most toxic substances known and the most convenient trigger material for bombs; and 3) the disposal of nuclear wastes. Three Mile Island is discussed in detail. This is a lucid and well-documented account for the nonspecialist.

606 Trager, Oliver. "Nuclear Energy," in Trager, Oliver, **Our Poisoned Planet: Can We Save It?** New York: Facts on File, 1989, p. 42-86.

This is a chapter in a book that contains editorials from American and Canadian newspapers. Topics include: a leak at a nuclear power plant in Oklahoma; Chernobyl; Rocky Flats; Hanford; and Three Mile Island. A wide variety of newspapers and viewpoints are represented. Each clipping is in the original typeface of the newspaper in which it appeared and is accompanied by the date it was published. The entries are in a loose chronological order. This unusual format makes for very interesting reading.

607 Turco, Richard P., Owen B. Toon, Thomas P. Ackerman, James B. Pollack, and Carl Sagan. "The Climatic Effects of Nuclear War," in *Scientific American*, v. 251, n. 2, August 1984, p. 33-43.

The article summarizes the consequences of nuclear war that could cause a global "nuclear winter." The belief is that the firestorms that develop after the detonation of many nuclear bombs will create enormous volumes of smoke and dust that move up into the atmosphere and out over a vast region. Sunlight could not penetrate through the haze, producing a decrease in temperature at the surface of the Earth. Similar scenarios have resulted from giant violent volcanic eruptions such as the one by Krakatau in 1883. Well illustrated and written, the results of the authors' computer simulations are terrifying.

608 Union of Concerned Scientists staff (eds.). **Safety Second: The NRC (Nuclear Regulatory Commission) and American's Nuclear Power Plants**. Bloomington, IN: Indiana University Press, 1987, 208p.

This is an analysis by the Union of Concerned Scientists of the NRC's operations between 1975 and 1985. The NRC is criticized on four counts: 1) avoiding safety issues; 2) limiting public input; 3) failing to enforce safety standards and conduct inspections; and 4) maintaining a relationship with the industry it was designed to regulate. Included in the discussion is a description of factors that led to the accident at Three Mile Island. The book concludes with recommendations to improve the commission. The introduction gives a good overview of the history and politics of the nuclear power industry. The footnotes and index are excellent. The text is dry but is suitable for the general reader.

609 Wagner, Richard H. "Radiation: Use and Abuse," in Wagner, Richard H., **Environment and Man**. New York: W. W. Norton, 1978, p. 311-335.

This is a chapter in a book that discusses man's impact on his natural environment. The chapter deals with all forms of radiation, which is defined as the "emission and transmission of energy as waves or particles." The discussion of microwaves and ionizing radiation is particularly good. Additional topics include atomic fission, radioactive fallout, contamination of the food chain, and radiation sickness. The author's point that the mutational rate of chromosomes is directly proportional to the total amount of radiation is well made.

610 Walker, Charles A., Leroy C. Gould, and Edward J. Woodhouse (eds.). **Too Hot to Handle?** New Haven, CT: Yale University Press, 1983, 209p.

The book has seven chapters, each written by a different expert, that discuss the issues surrounding the management of radioactive waste. Topics include: the radioactive waste management problem; the technology of radioactive waste management; health risks; public attitudes; acceptable risk; and the politics of radioactive waste management. References are at the end of each chapter. An index concludes the book. This is a scholarly treatment of an important subject.

611 Wasserman, Harvey, and Norman Solomon. **Killing Our Own**. New York: Delacorte Press, 1982, 368p.

The authors investigate the amount of atomic radiation in the United States and its health effects. Their research documents that the infant mortality rate near Three Mile Island jumped dramatically after the nuclear accident; that uranium waste is spilled into Lake Mead, which is a major water source for Los Angeles; that American GIs who were in Hiroshima and Nagasaki after the atomic bombs were dropped have cancer rates substantially higher than the normal population; and that bread in the United States contains plutonium released into the atmosphere during above-ground nuclear testing. This is a shocking account of the effects of an industry that proponents argue is only hypothetically dangerous. The introduction by noted child expert, Dr. Benjamin Spock, will alarm parents. Extensive notes and a good index accompany the text.

612 Wasserman, Harvey, and Norman Solomon. "X-Rays and the Radioactive Workplace," in **Killing Our Own**. New York: Delacorte Press, 1982, p. 125-139.

This is a chapter in a book that calculates the amount of atomic radiation in the United States and analyzes its health risks. The chapter begins by describing the discovery of x-rays by William Roentgen and their early application by the medical community. Side effects, detected shortly after the initial use of x-rays, also are examined. Mammography, radiation therapy, and the health effects of x-rays on the unborn child are discussed in detail. Extensive notes for the chapter are at the end of the book. This is an informative treatment suitable for everyone to read.

613 Whitman, Lawrence E. **Fire Safety in the Atomic Age**. Chicago: Nelson-Hall, 1980, 265p.

This book is a manual, written in nontechnical language, on how to establish an effective system to fight fires in radioactive materials. The author believes that the hazards are neither to be "feared nor ignored" but to be reconciled and lived with responsibly and safely. The book contains 17 chapters, the majority of which discuss general topics including the benefits of radiation, the hazards of radiation, radiation physics for the layperson, transporting nuclear waste, and nuclear weapons. Within each chapter, specific fire-fighting information relevant to the subject is given. A good index accompanies the text.

13

Land Mismanagement

Environmental degradation comes not just from air and water pollution, radioactive and toxic waste, and ozone depletion, but from the destruction of tropical rainforests, wilderness, and arable land. Though the threats posed by the loss of these habitats through mismanagement are not immediately discernable, they are real nonetheless. Once tropical forests burn to the ground and fertile land turns to desert, they and the wildlife they support are gone forever.

Loss of the tropical rainforests is perhaps the most publicized and serious example of land mismanagement. These woods once covered approximately 1.5 billion hectares in the equatorial belt but have been cleared in recent years for ranching, farming, timber, and development. Now only three large regions of tropical rainforest, comprising 900 million hectares, remain: the Amazon Basin in South America; the Congo Basin in west Africa; and the Malay-Indonesian peninsula and island chain in southeast Asia. Estimates for rates of deforestation range between 10 million and 20 million hectares per year. The Ivory Coast in west-central Africa has removed 66 percent of its forests since the late 1960s: Madagascar has cut 93 percent. The reasons for the destruction are economic. Most of the countries that have rainforests are poor and overpopulated. The governments view the woods as a hindrance to development, prefering to exploit them quickly rather than maintaining them as a sustainable and future resource. Cattle can be raised on the deforested land, for example, for export to the United States. The counterargument is that rainforests contain many important products upon which the industrialized world depends, including gums, resins, latexes, waxes, rubber, rattan, bamboo, dyes, tropical oils, steroids, and nuts.

The rainforests also benefit us in other important ways. At least 25 percent of medicinal substances were derived originally from plants found in the equatorial regions. Hopes run high that additional medicinal plants await discovery. More critically, however, the rainforests house two-thirds of the world's plant and animal species. Destruction of this biological diversity would disrupt delicate ecosystems and affect the overall balance of nature. Rainforests also buffer the temperature of the lower atmosphere, absorb rainfall and release moisture, and generate oxygen while removing carbon dioxide, a greenhouse gas, from the air.

Many researchers propose that the dawn of agriculture was the turning point for humans that pushed them on a path toward land mismanagement and environmental destruction. It allowed humans to manipulate natural balances by replacing complex ecosystems where numerous species coexisted with simple ones where one crop flourished. Furthermore, once humans had the ability to grow their own food, they began to multiply and to urbanize. When land was plentiful, the consequences of bad agricultural practices and development were never severe enough to warrant changing methods. Another piece of land was always available. This attitude led directly to the problems of vanishing wilderness, overcrowded cities, desertification, and extinction that threaten our world today.

Land mismanagement is a vast subject. The list of citations that follows is meant as a starting point for further investigation and is not a comprehensive review of the available literature. Reflecting current global concern, the chapter relies primarily on sources about vanishing wilderness and tropical rainforests. Excellent overviews include: *In the Rainforest* by Catherine Caulfield; *Vanishing Paradise: The Tropical Rainforest* by Stephen Dalton, George Bernard, and Andrew Mitchell; *Our Vanishing Wilderness* by Mary Louise Grossman, Shelly Grossman, and John N. Hamlet; *The Mighty Rainforest* by John Nichol; *Life Above the Jungle Floor* by Donald R. Perry; and *The Quiet Crisis* by Stewart L. Udall. Most of the volumes are beautifully illustrated. Other topics addressed in the chapter are land-use planning (note Gordon A. Bradley), the history of agriculture and its effects on ecosystems (note Robert J. Braidwood, the Conservation Foundation), urbanization (note William Bronson, Kingsley Davis, Thomas R. Detwyler and Melvin G. Marcus, *Scientific American*, Anne Whiston Spirn), travel to tropical rainforests (note James L. Castner, Adrian Forsyth and Kenneth Miyata), biological diversity and extinction (note the Conservation Foundation, Roger L. DiSilvestro, Paul Ehrlich and Anne Ehrlich, R. J. Hoage, Les Kaufman and Kenneth Mallory, Edward O. Wilson), desertification (note El-Baz Farouk and Mohammed H. A. Hassan, Paul B. Sears), wetlands and beaches (note Charles H. W. Foster, Wallace Kaufman and Orrin Pilkey), arid lands (note Carle Hodge and Peter C. Duisberg, Martin Simons, Gilbert F. White), land subsidence (note Thomas L. Holzer, A. I. Johnson and others), rangelands and cattle grazing (note Merton R. Love, William D. Rowley), development in the Amazon Basin (note Emilio F. Moran), endangered rivers (note Tim Palmer), the timber industry (note Robert Repetto, Gordon Robinson), and saving the redwoods (note Susan R. Schrepfer).

614 Beddington, John R., and Robert M. May. "The Harvesting of Interacting Species in a Natural Ecosystem," in *Scientific American*, v. 247, n. 5, November 1982, p. 62-69.

Written by two university professors, the article discusses the effects of harvesting krill—the primary food source for the baleen whale—in the southern oceans surrounding Antarctica. The emphasis is on the threat to the already overexploited baleen whale. A brief history of whaling begins the article. Excellent figures accompany the straightforward text.

615 Bradley, Gordon A. (ed.). **Land Use and Forest Resources in a Changing Environment**. Seattle, WA: University of Washington Press, 1984, 222p.

This collection of papers on land use planning was given initially at a national symposium on the urban/forest interface in Seattle, Washington, in October 1982. The emphasis was on managing land and resources in the era of a changing environment. Papers are broken into six categories: 1) the setting, which is an overview of the relationship between cities and forests; 2) the legal, political, and administrative aspects of managing public lands; 3) economics; 4) the physical and biological dimensions of cities and adjacent public space; 5) the social and psychological ramifications; and 6) recommended courses of action. Some figures accompany the text.

616 Braidwood, Robert J. "The Agricultural Revolution," in *Scientific American*, **Man and the Ecosphere**. San Francisco: W. H. Freeman, 1971, p. 17-25.

This article, originally published in *Scientific American* in September 1960, argues that the beginning of agricultural practices was a turning point in the evolution of human societies that propelled them on a path toward environmental destruction. The increased ability to grow and store food led to lower death rates and increased population, which stressed resources. In addition, cultivated land replaces complex ecosystems where many kinds of plants and animals flourish with simple ecosystems where a single crop or animal survives. Because nature requires diversity to thrive, this last pattern is particularly lethal to the environment. This is an interesting paper that details the beginnings of the agricultural revolution and the management of land nearly 10,000 years ago.

617 Bronson, William. **How to Kill a Golden State**. New York: Doubleday, 1968, 224p.

This is a pictorial essay that focuses on the urbanization of California in the 1950s and 1960s. The loss of wilderness lands and ecosystems is unfathomable. The tone is angry. The book is particularly interesting given the continued overdevelopment of California in the 1970s and 1980s.

618 Brower, Kenneth. "Losing Paradise," in **Wilderness**, v. 53, Winter 1989, p. 20-31.

The Hawaiian archipelago, the most isolated on Earth, is described as an example of an island ecosystem. Of the island's native species, 96 percent do not occur elsewhere. The isolation of the habitat makes it particularly fragile and susceptible to human interference. As a result, 40 percent of Hawaii's native plants are endangered or are candidates for endangered designation. The lack of conservation efforts is discussed also. The establishment of federal reserves is encouraged. The photographs are spectacular and complement the narrative well. This is recommended to anyone interested in conservation and island ecosystems.

619 Castner, James L. **Rainforests: A Guide to Research and Tourist Facilities at Selected Tropical Forest Sites in Central and South America**. Gainesville, FL: Feline Press, 1990, 380p.

The author, a biologist, has put together a travel guide for the person interested in visiting a rainforest. The book contains four parts: 1) countries, sites, and facilities, which discusses the specifics about Peru, Ecuador, French Guiana, Venezuela, Trinidad, Costa Rica, and Panama; 2) rainforest information sources; 3) hands-on organizations; and 4) sources of funding. Many photographs are included. This is a good source for the individual who wants to go to a remote locality in one of the countries mentioned.

620 Caulfield, Catherine. **In the Rainforest**. New York: Random House, 1985, 304p.

This is an eloquent history of the rainforest of the Amazon and its native people. The effects of colonization and industrialization are examined in detail. The destruction documented by the author is heartbreaking. This is an excellent look into a vanishing ecosystem and will enlighten those who read it.

621 Committee on Public Works, U.S. House of Representatives. **Oil Spillage—Flood Damage, State of California**. Washington, DC: U.S. Government Printing Office, 1969, 375p.

This volume is the transcript of the hearings before the Subcommittees on Flood Control and on Rivers and Harbors of the Committee on Public Works of the U.S. House of Representatives in February 1969, in California to assess the damage from major flooding in the central part of the state. The proceedings are dry and full of political asides, but are interesting because of the insight they give to the bureaucracy that surrounds any government hearing. The text is well illustrated with numerous maps, graphs, and tables that detail the extent of damage and the effect on soil conservation. The transcript is recommended reading for anyone interested in the federal government's response to state disasters.

622 The Conservation Foundation. "Agriculture and the Environment in a Changing World Economy," in The Conservation Foundation, **State of the Government: A View Toward the Nineties**. Washington, DC: The Conservation Foundation, 1987, p. 341-406.

This discussion of agriculture in the context of pollution, conservation, and human health risks is extremely informative and interesting. Part of a report by the Conservation Foundation designed to allow intelligent decisions to be made about environmental policy in the years to come, the chapter emphasizes the resource and environmental effects of changing agricultural practices in the United States and the rest of the world. Topics include surface-water pollution, soil productivity, groundwater quality, destruction of wildlife habitat, land use, irrigation, pesticides, surplus production, and policies for erodible cropland and wetland. A section entitled "Looking Ahead" concludes the discussion. Suggestions for further reading, a long reference list, and numerous figures accompany the text.

623 The Conservation Foundation. "Natural Resources: Land, Water, and Energy—Land," in The Conservation Foundation, **State of the Government: A View Toward the Nineties**. Washington, DC: The Conservation Foundation, 1987, p. 195-222.

This is a section in a chapter in a report written by the Conservation Foundation on existing and potential environmental policies, issues, and problems in the 1980s and 1990s. Topics include: land use; urban revitalization; growth management; cropland; rangeland and pastureland; and forestland. The emphasis is on the decrease in land used for agricultural purposes and the increase in land needed for urban development. The text is scholarly and well written. This is an excellent overview recommended to everyone.

624 The Conservation Foundation. "Protected Lands, Critical Areas, and Wildlife," in The Conservation Foundation, **State of the Government: A View Toward the Nineties**. Washington, DC: The Conservation Foundation, 1987, p. 275-340.

This chapter in the Conservation Foundation's report on the state of the environment focuses on natural resources that are not commercially viable: parks, wetlands, and wildlife. Included in the discussion are the national park system, national forests, the Bureau of Land Management, national wildlife refuges, wilderness areas, wild and scenic rivers, and national scenic and historic trails. State, local, and private parks also are described. Programs to protect game, threatened, and endangered species are summarized. Numerous figures, graphs, tables, and an extensive reference list accompany the text.

625 The Conservation Foundation. "Protection of Biological Diversity: The Challenges of a Broadening Perspective," in The Conservation Foundation, **State of the Government: A View Toward the Nineties**. Washington, DC: The Conservation Foundation, 1987, p. 535-576.

The chapter, in a report by the Conservation Foundation on environmental policies, practices, and problems in the 1980s and 1990s, begins with a discussion of biological diversity and the need for its protection. Reasons are given from agriculture, health and medicine, tourism and recreation, and ecology. The discussion continues with the enumeration of the threats and challenges to preserving biological diversity. The section entitled "Concluding Thoughts" is intelligent and insightful. Suggestions for further reading and a thorough reference list are included. This thoughtful and intellectual essay should be read by everyone who is not sure why the protection of our biological resources is necessary.

626 Dalton, Stephen, George Bernard, and Andrew Mitchell. **Vanishing Paradise: The Tropical Rainforest**. Woodstock, NY: Overlook Press, 1990, 176p.

This is an eloquent pictorial essay about the world's tropical rainforests. The text gives just enough information to complement the photographs. Chapter titles are "The Awakening Forest"; "The Atmosphere of the Forest"; "The Variety of Life"; "Communication"; "Movement"; "Finding Food"; "Hunting Prey"; "Camouflage"; "Reproduction"; "The Upper Canopy"; "Decay and Renewal"; and "What You Can Do." The rainforest is spectacular in its beauty and its uniqueness. It is amazing that not everyone is crying out to save it.

627 Davis, Kingsley. "The Urbanization of the Human Population," in *Scientific American*, **Man and the Ecosphere**. San Francisco: W. H. Freeman, 1971, p. 267-279.

The article, written by several researchers at Columbia University, was first published in *Scientific American* in September 1965. The authors document the trend of people moving from rural to urban areas, defined as cities of 100,000 or larger, and discuss the implications of the trend. The rate of urbanization increased sharply between 1950 and 1960 compared to the previous 50 years. The article treats nonindustrialized nations separately from industrialized countries. Remarkably, trends in the two groups are similar. Numerous graphs augment the text. This is an excellent introduction to trends in urbanization.

628 Detwyler, Thomas R., and Melvin G. Marcus. **Urbanization and Environment: The Physical Geography of the City**. North Scituate, MA: Duxbury Press, 1990, 237p.

This book describes the city as a unique ecosystem. Chapters cover topics such as ecology, geology, climate, water runoff, noise, plants and wildlife, air pollution, soils, and the natural hazards of cities. This is an intriguing look at the urban landscape from an unusual perspective.

629 DiSilvestro, Roger L. **The Endangered Kingdom: The Struggle to Save America's Wildlife**. Wiley Science Editions. New York: John Wiley, 1989, 241p.

The book attempts to discover the impact of the Endangered Species Act and the effectiveness of animal management programs in the United States. The author outlines the history of animal protection and wildlife management policies by telling the stories of the gray wolf, the grizzly bear, the pronghorn antelope, the California condor, the western diamondback rattlesnake, the bowhead whale, and bats, among others. The book has three parts: 1) game animals; 2) endangered species; and 3) nongame wildlife. A good bibliography and index are included.

630 Douglis, Carole. "A Corridor in Peril," in *Wilderness*, v. 53, Winter 1989, p. 32-37.

The author describes the lower Rio Grande Valley of Texas where an unusual coalition of environmentalists, farmers, business people, and government officials is wreaking havoc on the ecosystem. The area grows significant amounts of citrus and, therefore, is losing much of its original brushland. Because the region is economically depressed, land is cheap. As a result, the U.S. Fish and Wildlife Commission is buying land for conservation and the area is inundated by over 400,000 tourists each year. The protected land, however, is in several pieces, making it difficult, if not impossible, to preserve the original ecosystem. Meanwhile, developers continue to clear brushland to build more houses in the citrus-growing area. The photography is superb.

631 Ehrlich, Paul, and Anne Ehrlich. **Extinction: The Causes and Consequences of the Disappearance of Species**. New York: Random House, 1981, 305p.

The noted population biologists and environmental activists have written a well-researched and entertaining book about the interrelationships among the living things that define an ecosystem. They warn that not until it is too late will we know which species are critical to the survival of the planet. The analogy they use is that of a "rivet popper," a man who pries the rivets out of the wing of an airplane. "Don't worry," he says, " I've taken a lot of rivets out already and the wing hasn't fallen off yet!" The book is in four parts: 1) "Introduction"; 2) "Why Should We Care?"; 3) "How Are Species Endangered by Humanity?"; and 4) "What Are We Doing and What Can We Do?" The appendix is a taxonomy of the organisms discussed in the book. Extensive notes and an index are included.

632 Eisenberg, Evan. "Back to Eden," in *The Atlantic Monthly*, v. 264, November 1989, p. 57-89.

The article focuses on Wes Jackson and his argument that modern agricultural practices are destroying the topsoil and polluting the environment. In 1976, Jackson founded the Land Institute to study "alternatives in agriculture, energy, waste management, and shelter." He began work on a new kind of prairie capable of maintaining its original ecosystem and also of producing food for humans. The experiment, designed to take 100 years, is described thoroughly. The article is interesting and informative although somewhat long-winded.

633 Farouk, El-Baz, and Mohammed H.A. Hassan. **Physics of Desertification**. Dordrecht, Netherlands: Martinus Nijhoff, 1986, 473p.

Desertification is the process by which previously arable land becomes desertlike. Usually it is the result of mismanagement of the environment but it also has some natural causes. This volume contains papers presented at a workshop in 1980 at Trieste, Italy. Each paper discusses an example of desertification. Examples are varied in location—from the Sudan to Mars. Also included are papers that describe efforts to combat and prevent the phenomenon. The text is very detailed and may not be of interest to most people.

634 Ferguson, D., and N. Ferguson. **Sacred Cows at the Public Trough**. Bend, OR: Maverick Publications, 1983.

The book documents the effect of the livestock industry on the American West. Overgrazing, diversion of water, and irrigation are discussed. The cooperation of the federal government in the formulation of its policies also is detailed. The damage to the ecosystem is alarming.

635 Forsyth, Adrian, and Kenneth Miyata. **Tropical Nature: Life and Death in the Rainforests of Central and South America**. New York: Charles Scribner's Sons, 1987, 272p.

Written by two field biologists who worked extensively in the rainforests of Costa Rica and Ecuador, this book is a celebration of the diversity of tropical ecosystems. Each of 17 chapters discusses one aspect of the rainforest. Advice on how to visit the tropical world, what to bring, and what tour to choose is included. Many beautiful illustrations complement the upbeat and simple text.

636 Foster, Charles H. W. "Coastal Wetlands Protection Program in Massachusetts," in Woods, Barbara (ed.), **Eco-Solutions**. Cambridge, MA: Schenkman Publishing Company, 1972, p. 155-167.

The article summarizes the coastal wetlands protection program in Massachusetts beginning about 1954 when 45,000 acres of coastal wetlands were inventoried by the Department of Agriculture. Two factors are cited in the purposeful destruction of wetlands: public health and agricultural drainage. The legislative history of the program is outlined. Massachusetts general laws pertaining to coastal wetland destruction also are included. The citizen's letter of protest that appears at the end of the article is interesting because it presents an alternate view that the citizen maintains is equally environmentally sound as preservation.

637 French, Herbert E. **Love of Earth**. New York: G. P. Putnam's Sons, 1973, 314p.

This is an eloquent account of the beauty of the natural world around us. The author describes the land on which we live by exploring a rainforest, strolling on a beach, climbing a mountain, and descending into a cave. It is a true celebration of the land in its wildest form. A bibliography and index are included.

638 Goodland, Robert (ed.). **Race to Save the Tropics: Ecology & Economics for a Sustainable Future**. Washington, DC: Island Press, 1990, 218p.

This book emphasizes "applied tropical ecology" as a solution to the destruction of the rainforests. Topics include: the design of agricultural projects; the construction of dams; the integration of conservation with development; and the analysis of forest management. An introduction to applied ecology begins the book. A summary of environmental action opportunities, suggestions for how the citizen can save the rainforest, and a list of rainforest action groups conclude the book. An index and bibliography accompany the text.

639 Goudie, Andrew. **The Human Impact on the Natural Environment**. Cambridge, MA: MIT Press, 1986, 338p.

This is a basic text for an introductory course in environmental studies and land resource planning written by a Professor of Geography at Oxford University. Although the book is a source of general information, it is included in this section of the bibliography because over half the volume emphasizes people's impact on the land and the creatures that live on it. Chapters of interest are "The Human Impact on Vegetation"; "The Human Impact on the Soil"; "The Human Impact on Animals"; and "Human Agency in Geomorphology." The book has a thorough bibliography and a good index. Illustrations are numerous. This is an excellent and lucid introduction to how people have impacted the environment.

640 Grossman, Mary Louise, Shelly Grossman, and John N. Hamlet. **Our Vanishing Wilderness**. New York: Grosset & Dunlap, 1969, 324p.

This is a pictorial essay about the wilderness and wildlife that is disappearing quickly from the United States. Each chapter of the book focuses on one habitat, such as wave-swept and sheltered shores, watery lowlands, upland and northern forests, grasslands and sagebrush country, the desert, and new mountains. The photographs are beautiful and the text is straightforward. The front and back inside covers are maps of the areas visited by the authors. A bibliography and index are included.

641 Hall, Peter. "London," in Woods, Barbara, **Eco-Solutions**. Cambridge, MA: Schenkman, 1972, p. 201-222.

This essay describes the efforts of planners in London since World War II as they tried to develop a comprehensive plan to control unbridled growth. Various acts, including the Green Belt Act which provided open space and the Town and Country Planning Act, are described. The creation in 1965 of the Greater Council of London, which oversees every aspect of growth and development of the city, is also outlined. The essay gives a good overview of the problems of large metropolitan areas, using London as the specific example. The difficulties in defining a metropolitan area are detailed.

642 Hoage, R. J. (ed.). **Animal Extinctions: What Everyone Should Know**. Washington, DC: Smithsonian Institution Press, 1985, 192p.

The 12 papers in this volume were presented at the first National Zoological Park symposium for the public. The theme was extinction, the elimination of species and habitats. The contributors are university professors, government and nonprofit organization officials, and zoologists. Topics include: a general overview of extinctions; the vulnerability of species to extinction; extinction as a natural process; the present extinction spasm; animal and plant species used in medicine and industry; strategies for preservation in the wild and in captivity; sustainable exploitation of wildlife; and reintroduction of species into the wild. The papers are well written and argued and some are illustrated.

643 Hodge, Carle, and Peter C. Duisberg. **Aridity and Man—The Challenge of Arid Lands in the United States**. Washington, DC: American Association for the Advancement of Science, 1963, 584p.

This book addresses all aspects of arid land research including the historical framework, water and its use, minerals and energy sources, soils, and native plants and animals. Particularly intriguing are the chapters on human factors in desert development and the review of human history in arid regions. The last part of the book summarizes case histories from the southwestern United States. The case of Los Angeles in 1963 is especially enlightening given the city's rapid expansion in the 1970s and 1980s. The book is well illustrated and is suitable for anyone. The map of arid regions in the United States on the inside back cover is very informative.

644 Holzer, Thomas L. (ed.). **Man-Induced Land Subsidence** Reviews in Engineering Geology VI. Boulder, CO: Geological Society of America, 1984, 232p.

The volume contains data from a few spots where the economic impact of land subsidence is well documented. Nine papers are arranged in three categories: 1) "Fluid Withdrawal from Porous Media"; 2) "Drainage of Organic Soil"; and 3) "Collapse into Man-Made and Natural Cavities." The text is technical and dry. The volume is recommended to individuals with some scientific background.

645 Hultsman, John, Richard L. Cottrell, and Wendy Zales-Hultsman. **Planning Parks for People**. State College, PA: Venture, 1987, 310p.

This is an interesting volume that describes the work of a recreation specialist who plans, administers, and maintains parks. Chapters cover day use areas, campgrounds, evening programs in parks, trails, and specific use facilities such as those for off-road vehicles. The authors discuss the compromises that are necessary when weighing the benefits of a park against the impact on the environment. The illustrations complement the text well. The volume reads like it is designed to accompany an introductory class on the planning of parks.

646 International Association of Hydrological Sciences. **Challenges in African Hydrology and Water Resources**. Washington, DC: International Association of Hydrological Sciences, Publication no. 144, 1984, 587p.

Although this volume is quite technical and is primarily a study of the availability of groundwater, it is of use to the person interested in desertification. It elucidates the problems of water supply in desert areas, climate, and soil erosion—all problems exacerbated by overpopulation. A background in physical sciences will make this volume more enjoyable to the reader. Illustrations, graphs, tables, and citations are plentiful.

647 Iverson, Johannes. "Forest Clearance in the Stone Age," in *Scientific American*, **Man and the Ecosphere**. San Francisco: W. H. Freeman, 1971, p. 26-31.

First published in *Scientific American* in March 1956, this article is an interesting summary of the deforestation practiced by Stone Age humans. Evidence includes decreases in tree pollen in certain regions followed by increases in herbaceous plant pollen and in cereals and weeds. Stone Age humans probably used slash-and-burn techniques as demonstrated by the replacement of primeval forest with willow, aspen, and birch, of which birch flourishes in a mixed oak forest only after it has been burned. The validity of the theory is tested by archaeologists who cut down two acres of forest with little difficulty using Stone Age tools. The figures are excellent. This gives food for thought to people who maintain it is modern society that has led to environmental problems.

648 Jacobs, Jane. **The Death and Life of Great American Cities**. New York: Random House, 1965, 458p.

This is a somewhat dated volume on cities and their problems, including overcrowding, overdevelopment, politics, and pollution. The perspective of the 1960s is of historical interest. Written before the death of the inner cities characteristic of the 1980s, the book is only marginally relevant today.

649 Johnson, A. I., Laura Carbognin, and L. Ubertini. **Land Subsidence**. International Association of Hydrologists Publication no. 151. Washington DC: International Association of Hydrologists, 1984, 933p.

This volume is a compilation of papers presented at the Third International Symposium on Land Subsidence in Venice, Italy, in March 1984. The symposium was sponsored by the International Association of Hydrological Sciences, UNESCO, and the Italian Research Council. Most papers address natural forms of subsidence. Some discuss human-induced subsidence problems such as withdrawal of groundwater in mines or arid regions. One section emphasizes oil, gas, and mineral extraction; another, the economic, environmental, and legal issues of subsidence. The volume is quite technical. Most of the papers, however, can be understood without a solid background in either mathematics or the physical sciences.

650 Kaufman, Les, and Kenneth Mallory. **The Last Extinction**. Cambridge, MA: MIT Press, 1986, 208p.

This book has an apocalyptic message: We are in the midst of one of the worst mass extinctions in the history of life on Earth. The culprit is human. Habitat destruction through complex chains of cause and effect is believed the primary and most serious threat. Over-hunting and over-fishing are secondary. Case histories are cited from the impact of development projects in the Amazon rainforest, the United States, and Canada. The last part of the book emphasizes options for the future such as preservation of minimum viable population sizes in zoos. Although the volume is a collection of papers by distinguished scientists, it is designed for a wide audience and is easy to read.

651 Kaufman, Wallace, and Orrin Pilkey. **The Beaches Are Moving**. Garden City, NY: Anchor Press, 1979, 326p.

This is a study of the rise of sea level that has occurred for the past 15,000 years (and continues today) and its impact on coastlines and residential beach property. The authors explain the geological processes responsible for changes in sea level and outline the consequences of building fortifications to protect the beaches and alter the natural order of things. Two appendices contain a checklist for buying or building a beach home and suggestions for where to see beaches moving. References and index are thorough. This is an informative book that is easy to follow.

652 LaBastille, Anne. **Assignment: Wildlife**. New York: E. P. Dutton, 1980, 243p.

The author, an ecological consultant and recipient of the 1974 World Wildlife Fund, International, Gold Medal for Conservationist of the Year, has written an eloquent and passionate account of her travels and work in Central America, the Caribbean, and the Amazon Basin. The projects include: establishment of a volcanic national park in Panama; protecting the grebe, an endangered bird, in Guatemala; and studying the rainforest in Brazil. Her descriptions of the native cultures and their attitudes toward the world around them are excellent. Despite some of the setbacks, the author is optimistic throughout the book, as is revealed clearly in the last chapter on predictions for the future.

653 Love, R. Merton. "The Rangelands of the Western United States," in *Scientific American*, **Man and the Ecosphere**. San Francisco: W. H. Freeman, 1971, p. 228-235.

This article first appeared in *Scientific American* in February 1970. The author discusses range management philosophy from the perspective that human activities must help nature—if they did not the land could not support the population that it does. He presents what are, in his opinion, several misconceptions that govern the view that nature is something with which we ought not to interfere. He maintains that each thing people do to their environment "improves" nature. The opinion is not a widely shared one, particularly today. The photographs and figures are good, although the map of federally controlled land in the western United States is outated—Redwoods National Park, for example, is missing because it had not been established in 1970.

654 Lowdermilk, Walter C. "The Reclamation of a Man-made Desert," in *Scientific American*, **Man and the Ecosphere**. San Francisco: W. H. Freeman, 1971, p. 218-227.

This article, describing the reclamation project undertaken by the State of Israel to irrigate arid land that had once been more fertile, appeared initially in *Scientific American* in March 1960. The reduction in agricultural productivity in the land during the millenia that preceded the reclamation effort is believed the result of overgrazing and poor cultivation practices. Greater than three feet of topsoil had been eroded. The success of Israel at irrigating its soil is thought by the author to be able to be duplicated by other countries plagued by barren land. The editors of the volume in which the article appears, however, disagree.

655 Lowry, William P. "The Climate of Cities," in *Scientific American*, **Man and the Ecosphere**. San Francisco: W. H. Freeman, 1971, p. 180-188.

This article was first published in *Scientific American* in August 1967. The author discusses the reasons why cities are warmer than rural areas. These include: changes in surface material, i.e., concrete for grass; increased surface roughness, generation of heat, and air pollution. If urbanization occurs over large areas, immense changes in climate may take place over vast regions of the continents. The increase in temperature over cities is dramatically illustrated by a temperature profile over London and surrounding areas. The figures that accompany the text are excellent.

656 Mark, Shelly M. "Hawaii's Experience with Its State Land Use Law," in Woods, Barbara, **Eco-Solutions**. Cambridge, MA: Schenkman, 1972, p. 249-255.

Hawaii was the first state to introduce a statewide plan for land use. Implemented in 1961, the plan provides for uniform property assessment and tax collection and also has a statewide capital improvements program. The essay describes the land use plan and traces its successes and failures through its first 10 years. The author concludes that the plan is most effective in crisis situations and is unlikely to be adopted elsewhere without substantial modification.

657 McNeil, Mary. "Lateritic Soils," in *Scientific American*, **Man and the Ecosphere**. San Francisco: W. H. Freeman, 1971, p. 68-73.

The article describes the process of laterization in which mineral-rich earth of a certain composition turns into a hard rock when it is exposed to air.

The problem of laterization is especially acute in the tropics where lateritic soils become hard and impenetrable after cultivation, establishing a barrier to increased agricultural productivity. The soils are rich in iron and aluminum and occur generally between 30°N and 30°S latitude. Not a human-made environmental problem, laterization nevertheless illustrates the importance of understanding one's environment before attempting to change it.

658 Meijer, Willem. "Saving the World's Largest Flower," in *National Geographic*, v. 168, n. 1, July 1985, p. 136-140.

This is an interesting article about a fascinating flower, *Rafflesia arnoldii*, that grows in Malaysia and Indonesia to a size of about a yard across. First discovered in Sumatra in 1818 by a surgeon-naturalist, Dr. Joseph Arnold, the flower is now in danger of disappearing because of the destruction of the rainforests. The pictures are excellent.

659 Milne, Lorus J., and Margery Milne. **Ecology Out of Joint**. New York: Charles Scribner's Sons, 1977, 304p.

The authors use case histories to analyze the effects that people have had on the environment. The emphasis is on plants and animals that invade a new ecosystem. Examples are palm trees in Scotland; an American waterweed in Europe; channeled whelks from Cape Cod in San Francisco Bay; and African snails in Malaysia. Endangered species and the effects of chemical pollutants on wildlife also are discussed. An index and a section of notes are included.

660 Mitchell, Andrew W. **The Enchanted Canopy: A Journey of Discovery to the Last Unexplored Frontier, the Roof of the World's Rainforests**. New York: Macmillan, 1987, 255p.

This is an attractive and readable book about life 100 to 200 feet above the jungle floor written by a British naturalist and television producer. The author travelled around the world to compile this nontechnical and comprehensive survey of the biology of the canopy ecosystem. One interesting section describes the methods used to investigate the area above the ground: helicopters (too noisy and drafty); aerial walkways; scaffolds; climbing ropes; and hot air balloons.

661 Moran, Emilio F. **Developing the Amazon**. Bloomington, IN: Indiana University Press, 1981, 292p.

The book describes the impact of the opening of the TransAmazon Highway in 1970 on the world's largest rainforest, specifically the effectiveness of the early homesteading programs to develop the tropics for intensive agriculture. The introductory chapters on the problems and potential of a large rainforest and on ecosystems within the rainforest are valuable. The figures are informative and numerous. The bibliography is exhaustive. This is a good source for information about the ecology of the Amazon Basin.

662 Nichol, John. **The Mighty Rainforest**. London: David & Charles, 1990, 200p.

This is a beautifully illustrated and simply written introduction to the rainforests of the world for the nonspecialist. The ecosystem is described and the problems that threaten it are outlined. Chapters include: "Rainforests of the World"; "The People of the Rainforest"; "What Are the Alternatives?"; "What Is a Rainforest?"; "Why Are Rainforests in Danger?"; "What Is Being Done to Help"?; "Animals of the Rainforest"; "What Happens When Rainforests Disappear?"; and "Organisations Working in Rainforest Conservation." A bibliography and index round out the volume. The inside front and back covers are maps that show areas of rainforest throughout the world.

663 Palmer, Tim. **Endangered Rivers and the Conservation Movement**. Berkeley, CA: University of California Press, 1986, 316p.

The book, a project of the River Conservation Fund in Washington, D.C., tells the history of river conservation in America. The author describes the first battle in the river conservation movement: a campaign by John Muir to prevent the creation of a reservoir in the Hetch Hetchy Valley of the Tuolumne River in the Sierra Nevada. The campaign was unsuccessful. Since that time, however, other rivers have been declared as wild and scenic. Many of the events surrounding the establishment of the National Wild and Scenic Rivers System are detailed. The book includes a list of endangered rivers, a section of beautiful color photographs, an index, and extensive notes.

664 Passow, Shirley S. "Land Reserves and Teamwork in Planning Stockholm," in Woods, Barbara, **Eco-Solutions**. Cambridge, MA: Schenkman, 1972, p. 227-247.

The development of Stockholm, Sweden, is described with emphasis on the city's policy of obtaining outlying public lands far in advance of when the land will be necessary for the expansion of the city. This policy has resulted in the creation of new living communities around the city of Stockholm with little financial burden on the municipal government or the taxpayer. The difficulty in implementing such a program in the United States is discussed. The primary problem in the United States is that outlying areas are rarely controlled by cities when the cities need it. Purchasing land many years in advance, however, also is impossible when most American cities are in financial distress and land costs are prohibitive.

665 Perry, Donald R. "The Canopy of the Tropical Rainforest," in *Scientific American*, v. 251, n. 5, November 1984, p. 138-147.

One estimate is that 40 percent of the world's species live in the treetops of the rainforest, approximately 100 to 200 feet above the ground. The article emphasizes the general ecosystem of the canopy, which the authors studied using an ingenious method of ropes and pulleys. The number of species that live their entire lives above the ground is staggering. This is a great introduction for the general reader to the ecology and importance of the rainforest.

666 Perry, Donald R. **Life Above the Jungle Floor**. New York: Simon & Schuster, 1987, 170p.

Dr. Perry is an American biologist who has engineered a "canopy research system" to allow investigation of large tracts of rainforest above the jungle floor. His primary field area is Costa Rica where the apparatus will be anchored to high rock ridges above a valley. Overcoming the "obstacle of gravitation," as naturalist William Beebe wrote in the early twentieth century, will be a critical breakthrough in the study of the rainforest ecosystem. It has been for Dr. Perry. This is a beautiful and readable book.

667 Repetto, Robert. "Deforestation in the Tropics," in *Scientific American*, v. 262, n. 4, April 1990, p. 36-42.

Tropical rainforests are being cut at the rate of many thousands of square miles per year. The author argues convincingly that the extensive logging and burning of the forests are largely the result of government policies driven by the debt-burdened economies of the underdeveloped countries in which the forests grow. He further suggests that the economic hardships are exacerbated by the financial institutions in the developed countries. Numerous examples are cited. International cooperation is promoted as a possible solution. Excellent figures accompany the well-written text.

668 Repetto, Robert. **The Forest for the Trees? Government Policies and the Misuse of Forest Resources**. Washington, DC: World Resources Institute, 1988, 120p.

The author discusses government policies and subsidies toward the timber industry. He concludes that what is designed to encourage development of forests is, in fact, aiding their destruction. Twenty-seven million acres of forest land are eliminated each year in the tropics alone. This is particularly frightening when studies show the loss of trees in the warm latitudes of the Earth directly correlates with increased greenhouse gases.

669 Robinson, Gordon. **The Forest and the Trees: A Guide to Excellent Forestry**. Washington, DC: Island Press, 1988, 257p.

This guide is an excellent source for individuals interested in learning about proper and improper forest management. The book is divided into four parts: 1) the history of forestry in the United States; 2) the philosophy and science of multiple-use; 3) forest planning; and 4) supporting research and informed opinions. The last part is a listing of quotes necessary to document reports on forestry—it allows an individual to cite facts and opinions without visiting a forestry library or sending for pamphlets.

670 Romme, William H. and Don G. Despain. "The Yellowstone Fires," in *Scientific American*, v. 261, n. 5, November 1989, p. 37-46.

The authors address the huge fires that raged through Yellowstone National Park during the summer of 1988. More acres were destroyed in one 24-hour period during the fire than had burned in any decade since 1872. Officials of the park chose to let the fire burn when it began. It is this policy of allowing fires to burn that is under scrutiny in the article. By studying the ecology of Yellowstone for many years before the fire and for one year after, the authors established a five-stage succession for the growth of the forests. The article is well written and informative. The figures are superb.

671 Rowley, William D. **U.S. Forest Service Grazing and Rangelands: A History**. College Station, TX: Texas A&M University Press, 1985, 270p.

This is an account of the history of grazing on lands in the public domain of the American West. In the beginning, cattle and sheep were allowed to graze without limit or control by the federal government. This led to overcrowded and overgrazed rangeland. Damage, in some cases, was irreversible. Shortly after 1900 when the situation became critical, Congress designated the U.S. Forest Service as the control agency for grazing. The policy of the Forest Service is to provide public lands to many users from the individual who runs a family ranch to a corporation that owns many ranches. This is a thorough history of the grazing problem. Extensive notes, bibliography, index, and photographs accompany the text.

672 Schrepfer, Susan R. **The Fight to Save the Redwoods**. Madison, WI: University of Wisconsin Press, 1983, 338p.

The author traces the history of preservation efforts for the redwoods of northern California from 1917 to 1978 by interviewing some of the key participants and from examining their personal papers. Events are tied to political, economic, and intellectual trends of the times. Within this period, Redwood National Park was created and the Sierra Club became a powerful voice for wilderness advocates. The formation and growth of the Save-the-Redwoods League also is described. The book serves as an informative and practical guide to the preservation controversy. It is recommended to anyone interested in the environmental movement and the logging industry. The bibliography lists primarily unpublished sources.

673 *Scientific American*, **Cities: Their Origin, Growth and Human Impact**. San Francisco: W. H. Freeman, 1973, 297p.

This book contains articles that were first published in *Scientific American* between 1954 and 1973. The articles focus on how cities originated, evolved, and function in the modern environment. The volume contains five parts: 1) "The Earliest Cities"; 2) "Population, Health and the City Environment"; 3) "Urban Transport and City Planning"; 4) "Cities of the Developing World"; and 5) "Group Relations in Cities." Part 2 is the most interesting. Articles in this section include lead poisoning, air pollution and public health, and the climate of cities. Each article is written by an expert in the field. All are well illustrated.

674 Sears, Paul B. **Deserts on the March**. Washington, DC: Island Press, 1935, 241p.

This book was written in 1935 when much of the United States was in the midst of a great drought and the plains states were transformed into the Dust Bowl. The author, experiencing the desiccation, urged Americans to be mindful of the environment and to understand ecological relationships. The book had a tremendous impact and led to strict controls on soil erosion through proper land management as well as to increased concern for the overall environment. The author was one of the earliest ecologists and wrote a fascinating, lyrical book.

675 Sidle, R. C., A. J. Pearce, and C. L. O'Loughlin. **Hillslope Stability and Land Use**. Washington, DC: American Geophysical Union, 1985, 140p.

Although landslides are viewed as a natural phenomenon, most destruction occurs in rapidly urbanizing areas such as southern California. Slope failure also takes place in rural areas and this is the topic on which this volume focuses. Example are drawn from northern California, the Pacific Northwest, and New Zealand. The first chapters review the environments most likely to fail. Key natural factors such as slope gradient, moisture content of the soil, and so on are well illustrated. Other sections discuss the impact of logging, farming, slash-and-burn agriculture, road construction, residential development, and mining on slope stability. The glossary and reference list are excellent.

676 Simons, Martin. **Deserts: The Problem of Water in Arid Lands**. Oxford, England: Oxford University Press, 1967, 335p.

This is a beautifully illustrated text that describes the role of water in the desert or semi-desert land that covers nearly one-third of the Earth. Desert climate, plants, animals, and sources of water are discussed in separate chapters after the initial chapter outlining the problem of water in arid lands. One appendix describes Koppens classification of desert climates and provides a worksheet for readers to evaluate climates. Examples are given. The glossary is useful although quite short. This is an enjoyable book for anyone interested in deserts and water.

677 Spirn, Anne Whiston. **The Granite Garden: Urban Nature and Human Design**. New York: Basic Books, 1984, 290p.

The environmental hazards of the modern city and the creation of more human cities are discussed in this book. Included are sections on air pollution and urban smog, heat in the city, contaminated land, scarcity of potable water, the absence of vegetation, and waste removal. The author also provides solutions to some of the problems such as designing parks and open space; building water conservation systems; and planning future development. The beautifully written text is accompanied by numerous figures, extensive notes, and an index. It is an excellent reference.

678 Stern, Alissa J. "The Case of the Environmental Impasse," in *Harvard Business Review*, May-June 1991, p. 14-29.

A case study of the Vermilion Paper Company and its attempts to become more "socially responsible" is examined in this article. Also discussed are the company's dealings with its "offshore" enterprises in Equitania where it pursued an "environmentally sound" policy that promoted farming of essential tropical woods. After years of negotiations with the local government, Vermilion Paper was targeted by the Forest Defense Legion for a worldwide boycott because of its aim to log the rainforest. Four experts are interviewed about whether Vermilion should fight the inaccurate accusations or back off.

679 Swanson, R. H., P. Y. Bernier, and P. D. Woodward. **Forest Hydrology and Watershed Management**. International Association of Hydrological Sciences Publication no. 167. Washington, DC: International Association of Hydrological Sciences, 1987, 625p.

This is the proceedings of an international symposium in Vancouver, Canada, in August 1987 sponsored by the International Association of Hydrological Sciences and the Society of American Foresters. The volume includes papers that describe the hydrology of the entire forest watershed from erosion and sedimentation to acid precipitation. Numerous papers discuss the effects of clearcutting. The goal of the symposium was to focus attention on the problems of land management in environmentally sensitive forest areas, and it is with this goal in mind that the papers were solicited. The text is technical. The reader should have a familiarity with high school chemistry.

680 Udall, Stewart L. **The Quiet Crisis**. New York: Holt, Rinehart & Winston, 1963, 209p.

The author, a former secretary of the interior and a passionate environmentalist, has written a lyrical account of the destruction of wilderness lands in the United States. Conservation issues are introduced by outlining the actions of the men who championed them. John Muir, for example, is discussed in the chapter on wild and park lands. Several color plates of pristine wilderness and urban development are included. The introduction is by President John F. Kennedy. This is an eloquent treatment of an important problem that was not yet recognized in the early 1960s.

681 Vining, Daniel R., Jr. "The Growth of Core Regions in the Third World," in *Scientific American*, v. 252, n. 4, April 1985, p. 42-49.

The author details the trend in Third World countries toward increased populations in the cores of their cities. In South Korea, for example, 36 percent of all citizens lived in or near Seoul in 1980 whereas only 18 percent did in 1955. This creates problems of stability for the nations as the rapid growth places demands on housing, transportation, and sanitation that some governments cannot afford. Crowds and pollution are so extreme in several cities of the Third World that experts feel the carrying capacity has been reached.

682 Voll, Sarah Potts. **A Plough in the Field Arable: Western Agribusiness in Third World Agriculture**. Hanover, NH: University of New Hampshire Press, 1980, 213p.

A series of agricultural projects sponsored by private Western enterprises in the underdeveloped nations of Africa and the Middle East is discussed in this book. The volume takes both an historical and a modern view, covering projects over the last 50 years. The first chapter is a review of the information available on multinational enterprises, technological transfer, and agricultural development. The second chapter provides an overview of the general problems inherent in an agricultural project and the specific issues raised when a foreign company is involved. The remaining chapters focus on individual projects from pre-independent Africa days to the modern era. The book is well written and easy to read. References are at the end of each chapter and project discussion.

683 White, Gilbert F. (ed.). **The Future of Arid Lands**. Washington, DC: American Association for the Advancement of Science, 1956, 453p.

This volume is a collection of papers and recommendations that grew from an International Arid Lands Meeting in the spring of 1955 for the maintenance of the world's arid lands. One emphasis of the meeting was to prevent the destruction of semi-arid land by social and economic policies that prevent the implementation of new agricultural techniques. Papers discuss the predictability of the water supply, the use of resources, and the adaptation of plants and animals to arid conditions. Some papers are well illustrated, others are not. The text is not technical and is suitable for the general reader.

684 Wilson, Edward O. "Threats to Biodiversity," in *Scientific American*, v. 261, n. 9, September 1989, p. 108-116.

The author is a recipient of the Tyler Award for environmental achievement and is a professor at Harvard University. The article discusses the destruction of habitats in the tropics by human activity and the resultant extinction of thousands of species. Today, biodiversity is at its lowest levels since 65 million years ago at the end of the Mesozoic era when the dinosaurs disappeared. The length of time for a species to become extinct once its habitat is reduced or destroyed is a function of the size of the habitat but is commonly less than 50 years. This is a sobering and informative article and the figures are superb.

685 Woodwell, George M. "The Energy Cycle of the Biosphere," in *Scientific American*, **The Biosphere**. San Francisco: W. H. Freeman, 1970, p. 25-36.

This article was published in *Scientific American* in September 1970. The relationship among solar energy, photosynthesis, and the biosphere is discussed. The two types of food chains, the grazing or browsing food chain and the decay food chain, are described. The author also illustrates the effects of agriculture by comparing the food chains of an intact natural forest ecosystem, a degraded ecosystem, and an agricultural ecosystem. The article is well written. Numerous graphs and figures accompany the text.

14

Noise Pollution, Visual Pollution, and Pollution in Outer Space

Not all types of pollution involve noxious substances or contaminated resources. Some simply diminish our enjoyment of our surroundings. Two examples are noise pollution and visual pollution. Although noise is transient, dissipating when the offending sound is removed, the adverse effects of sudden and loud noise, such as hearing loss and psychological trauma, can be permanent. In contrast, visual pollution does not produce health risks. The issue is one of aesthetics. As urbanization and development devour the countryside, do people want their views of pristine wilderness obscured by billboards and neon signs that eventually break down and litter the landscape? For that matter, do people want to be bombarded with signs within the already chaotic and crowded urban environment?

An additional type of pollution is one with which we have no experience: the pollution of outer space. It is a problem that will become increasingly important as more nations launch rockets and the satellites already in orbit decay. Currently, outer space is beyond the jurisdiction of any nation. No rules govern how the area is used or misused. Any country could choose to dispose of their hazardous waste above the Earth's atmosphere with no legal ramifications, despite possible environmental consequences about which we have no knowledge.

Because these topics are removed from the immediate concerns of individuals interested in the environment, few books exist. The following is a compilation of most of the literature available with the exception of some highly technical textbooks used in graduate-level courses at universities.

686 Armour, Claude A. "Noise Abatement—Memphis Style," in Woods, Barbara, **Eco-Solutions**. Cambridge, MA: Schenkman, 1972, p. 173-181.

Noise is defined as "unwanted sound" and is everywhere in urban America. Two solutions have been suggested: reduction of noise at its source or replacement of noise with a different type of sound. This article discusses the regulation of noise in an urban environment, specifically in Memphis, Tennessee, which is believed by many to have the model noise control program. Ordinances and regulations, however, are similar to those in other cities where noise abatement has been slow. Education and participation of the public are considered the key ingredients in Memphis where value is placed on a quiet environment.

687 Beranek, Leo L. "Noise," in *Scientific American*, **Cities: Their Origin, Growth, and Human Impact**. San Francisco: W. H. Freeman, 1973, p. 158-167.
This article, which first appeared in *Scientific American* in December 1966, summarizes evidence for damage to the ear caused by noise in urban areas. The frequency (thousands of cycles per second) of noise is found to be the most important. In addition to hearing loss, psychological effects are discussed. The majority of the article, however, focuses on noise control. Suggestions include quieter buildings, motor vehicles, and airplanes. The text is well written but is outdated. Several informative graphs, including one depicting noise intensities of common household objects, accompany the text.

688 Bragdon, Clifford R. **Noise Pollution: A Guide to Information Sources**. Detroit: Gale Research, 1974, 524p.
This is a good bibliography that includes sources on the following topics: physiological effects; behavioral effects; community noise; environmental impact; acoustics; and education. Some of the entries are not readily available and readers are warned to keep this in mind as they search. The book, however, is a suitable first step in researching noise pollution.

689 Bronson, William. "Ear Pollution," in Hart, John (ed.). **The New Book of California Tomorrow**. Los Altos, CA: William Kaufmann, 1984, p. 71-74.
This essay was first written in 1967. The author complains of noise levels in the cities and on the highways of California. He states that greater than 6 million people in the U.S. suffer serious hearing loss of which noise is a major cause. The decibel scale is explained and the actions of the California legislature are condemned. The description of driving on the Oakland Bay Bridge is quite vivid. This is a passionate essay that is still valid today.

690 Bronson, William. "The Nervous Sign," in Hart, John (ed.), **The New Book of California Tomorrow**. Los Altos, CA: William Kaufmann, 1984, p. 66-68.
This is an amusing essay on the horrors of the billboards and the brightly lit signs that litter the roadsides of California. The author calls these collectively the Nervous Sign. He points out that most owners cease to wonder about their signs. As a consequence, bulbs burn out and aren't replaced. The sign is merely a symptom of the urbanization of California and the development of the mini-mall. The author urges sign control and points out that the towns of Menlo Park, Carmel, La Jolla, Santa Barbara, and Palm Springs have it and suffer no loss of profit.

691 Brown, Seyom, Nina W. Cornell, Larry L. Fabian, and Edith Brown Weiss. **Regimes for the Ocean, Outer Space, and Weather**. Washington, DC: The Brookings Institution, 1977, 257p.

This book was the result of the Technology and International Institutions Project of the Brookings Foreign Policy Studies Program conducted from 1972 to 1976. The study concentrates on the management of three areas that are outside the jurisdiction of any nation—the ocean, outer space, and the weather. Questions posed include: What criteria determine how the three areas are used? Who should decide the rules? and Who should implement them? They are answered in the context of U.S. foreign policy. In the discussion of these issues, much information is provided on hazards and pollution in the oceans and in outer space, and their effects on climate and the weather. The book is fascinating reading for anyone interested in the subject.

692 Cloud, Stan, and Ray March. "The Great Billboard Sellout," in Hart, John (ed.), **The New Book of California Tomorrow**. Los Altos, CA: William Kaufmann, 1984, p. 69-71.

This essay was first written in 1967. The authors discuss the Highway Beautification Act signed by president of the United States on October 22, 1965. The act was designed to eliminate the nearly 900,000 billboards on the nation's interstates. Unfortunately, the act contained provisions to ensure compensation to owners of billboards if their signs were required by law to be moved. In the years immediately following the legislation, federal funds paid owners and billboards came down rapidly. Federal money, however, soon disappeared and the remaining billboards stayed where they were.

693 Little, Ross A. "The Motor Vehicle Noise Problem and What Is Being Done about It in California," in Woods, Barbara (ed.), **Eco-Solutions**. Cambridge, MA: Schenkman, 1972, p. 183-195.

The author is a member of the California Highway Patrol. He describes how California reacted to the problem of controlling noise on the highways of the state. California has worked with automobile manufacturers to develop methods to measure noise levels and has enacted strong legislation that allows for strict enforcement. Figures that illustrate noise levels for cars on city streets, country roads, and freeways are included. Tables document the changes in noise levels as speeds change.

694 Organisation for Economic Co-operation and Development. **Fighting Noise: Strengthening Noise Abatement Policies**. Paris: OECD, 1986, 145p.

Between 1982 and 1985, members of the ad hoc committee on Noise Abatement Policies of the Organisation for Economic Co-operation and Development worked on methods to alleviate noise pollution. This report is a summary of the committee's findings. Topics include: the present situation of noise pollution; the outlook for the future; the adverse effects of noise; policies from 1970-1985; the cost of noise abatement; economic and energy constraints on noise abatement; and future policies toward noise pollution. The text is dry and is recommended to the individual who needs facts and figures on noise levels.

15

Health Problems in the Environment

Pollution and environmental degradation are not the only consequences of a technologically advanced society. As industries use more synthetic materials and complicated processes to manufacture goods, the less able they are to assess the possible hazardous effects of their practices. By-products may interact with other substances in the air, water, or human body to form dangerous toxins. The health risk to the global population, therefore, may be increased.

The works in this chapter concentrate on environmental health hazards. Some sources of problems are obvious: nuclear radiation (note David W. Lillie, Eric Chivian and others, Cyril L. Comar, and Donald B. Barlett and James B. Steele); x-rays (note Charles Panati and Michael Hudson, John W. Gofman and Egan O'Connor); indoor air pollution (note Birgitta Berglund); smog (note Walsh McDermott, Philip E. Graves and Ronald J. Krumm); toxic waste; asbestos (note Catherine H. Skinner and others, George A. Peters and Barbara J. Peters, Paul Brodeur); water pollution (note Allen E. Banik and Carlson Wade); and pesticides (note W. H. Hallenbeck and K. M. Cunningham-Burns). Others are not as apparent: ozone, which accumulates close to the ground giving rise to respiratory difficulties (note Jack Fishman and Robert Kalish); the affluent diet, which contributes to heart disease (note Erik P. Eckholm); power lines, which emit low-frequency electromagnetic radiation that may cause brain cancer and leukemia in children who live close to the high-tension wires (note David W. Lillie, Charles Panati and Michael Hudson); and silver dental fillings, which contain as much as 40 percent mercury, a highly toxic metal (note Sam Ziff). A few subjects are controversial. Recent investigations have suggested that dioxin, otherwise known as Agent Orange and responsible for the Love Canal incident, may not be as hazardous as was once suspected (note Fred H. Tschirley, Michael Gough). This information was heralded by the chemical industry and used as justification for the manufacture of other potentially toxic substances. Clearly, additional testing is necessary before abandoning all caution regarding this and other possibly lethal chemicals. Other topics include: the dangers of office work (note Jeanne Stellman and Mary Sue Henifin); the effects of climate on health (note William A. R. Thomson, Julius Fast); infectious drug resistance (note Tsutomu Watanabe); plagues and diseases (note W. F. Loomis, William L. Langer); low-level ionizing radiation such as that produced by radioactive isotopes in our own bodies (note Arthur C. Upton); video display terminals (note Elizabeth A. Scalet, Edward G. Godnig and John S. Hacunda); smoking (note E. Cuyler Hammond); rock dust (note Martin Cherniack); DES, an aid to pregnancy used between 1940 and 1970 (note Roberta J. Apfel and Susan M. Fisher); and agricultural practices (note Claude Aubert). Good overviews are: *The Picture of Health* by Erik P. Eckholm; *Public Health and the Environment: The United States Experience* edited by Michael R. Greenberg; *Environmental Hazards to Young Children* by Dorothy Noyes Kane; *The Health Detective's Handbook: A Guide to the Investigation of Environmental Health Hazards by Nonprofessionals* by M. Legator, B. Harper, and M. Scott; *Of Mice and Molecules* by Eric Skjei and M. Donald Whorton; and *Toxic Terror* by Dr. Elizabeth Whelan.

695 Apfel, Roberta J., and Susan M. Fisher. **To Do No Harm**. New Haven, CT: Yale University Press, 1984, 199p.

The authors, two doctors, tell the story of the use of DES from 1940 to 1970 to help mothers carry their babies to full term. In 1971, DES, a synthetic estrogen hormone, was found to cause infertility and cancer in the daughters of the DES mothers. The introduction, testing, and popularization of DES are the main focus of the book. The current state of knowledge of DES and its side effects also are summarized. The authors use the DES example to illustrate how technological advances in the medical community can have tragic consequences. Blame is not placed on anyone, but, rather, caution is urged to both patients and doctors.

696 Aubert, Claude. **Hunger and Health: Eleven Key Questions on Farming, Food and Health in the Third World**. Emmaus, PA: Rodale Press, 1985, 184p.

The author wrote the book to focus attention on the drawbacks of modern food technology and the benefits of autonomous food production, including less pollution and less impact on ecosystems. The text is organized under 11 questions. Examples are: Can agricultural practices be increased without using chemical fertilizers and pesticides? Do chemicals used in agriculture affect health? and Can irrigation programs be developed that are not harmful to health? Suggestions for further reading follow each question. The bibliography is extensive.

697 Banik, Allen E., and Carlson Wade. **Your Water and Your Health** (Revised 1990 Edition). New Canaan, CT: Keats, 1990, 140p.

This book contains most of the information the general reader wants to know about drinking water. Topics include: pollution in drinking water; the pros and cons of pure water; water hardness and your health; chlorine and your drinking water; fluorides and drinking water; and the effect of pure water on certain illness such as arthritis and diabetes. The authors are advocates of drinking much pure water (two to four glasses an hour!) to guarantee a longer and healthier life.

698 Barlett, Donald B., and James B. Steele. "Radiation, Waste, and the National Health," in Barlett, Donald B., and James B. Steele, **Forevermore: Nuclear Waste in America**. New York: W. W. Norton, 1985, p. 297-328.

This chapter in a book by two investigative reporters focuses on radiation-related diseases. Topics include: the effects of low-level radiation; radiation and the unborn; disregard for danger; workers at Hanford; federal safety standards; bad statistics; and the threat of unburied nuclear waste. The past behavior of the U.S. government regarding radiation exposure of the public does not bode well for the future of the nation's health. An easy-to-read chapter, this is suitable for anyone.

699 Berglund, Birgitta (ed.). **Recent Advances in Health Sciences and Technology**. Stockholm: Swedish Council for Building Research, 1984, 284p.

This volume contains the proceedings of the Third International Conference on Indoor Air Quality and Climate. As such, the full spectrum of viewpoints on the problem of indoor pollution is represented. Topics include: lung cancer risks from radon and passive smoke exposure; thermal discomfort; formaldehyde degassing; indoor air quality; and modern building techniques. Some of the papers can be quite technical. The volume, however, is a good source of information on an important subject.

700 Brodeur, Paul. **Outrageous Misconduct**. New York: Pantheon Books, 1985, 374p.

This account of the asbestos industry by a writer for *The New Yorker* focuses on the bankruptcy proceedings of the Manville Corporation undertaken by the company to avoid lawsuits brought against it by victims of asbestos-related illness. The author reveals that Manville and other companies successfully covered up the health problems caused by asbestos for 50 years before a few trial lawyers sifted through the evidence and discovered the hazards. The volume updates a series of four articles that originally appeared in *The New Yorker* magazine. The expose is interesting, but somewhat long-winded.

701 Cherniack, Martin. **The Hawk's Nest Incident: America's Worst Industrial Disaster**. New Haven, CT: Yale University Press, 1986, 184p.

The incident to which the author refers is the building of the Hawk's Nest Tunnel by the Union Carbide Corporation to transport water from a river to a power plant. Of the 5,000 men who worked on the tunnel, more than 10 percent died and 30 percent became ill from silicosis, a lung disease acquired by breathing rock dust. The author attempts to discover whether the company avoided safety precautions in the interest of speed and concludes that it did. This is a fascinating account of a grisly event.

702 Chivian, Eric, Susanna Chivian, Robert Jay Lifton, and John E. Mack (eds.). **Last Aid: The Medical Dimensions of Nuclear War**. New York: W. H. Freeman, 1982, 338p.

This is a collection of 20 papers about the effects of nuclear war on human health written for a general audience. Compiled by International Physicians for the Prevention of Nuclear War, Inc., the volume is horrifying. Papers are grouped into five sections: 1) introduction, which describes the general effects of a nuclear explosion; 2) nuclear war, 1945; 3) nuclear war, 1980s: the physical and medical consequences; 4) nuclear war, 1980s: the medical response; and 5) nuclear war, 1980s: environmental and psychological consequences. Suggestions for further reading and an index are included.

703 Comar, Cyril L. "Biological Aspects of Nuclear Weapons," in Bresler, Jack B. (ed.), **Human Ecology, Collected Readings**. Reading, MA: Addison-Wesley, 1966, p. 363-377.

The material in this chapter was given initially as a lecture on nuclear peril and disarmament. The author paints a balanced picture of the biological problems of radiation but oversimplifies and omits details. Problems derive from fallout and contamination and include either genetic or somatic effects. "Threshold" levels of exposure are discussed. Low and high levels of radioactive exposure are evaluated. Contributions of medical procedures also are estimated and are assumed equivalent to the background levels of radiation on the Earth. Some solutions for highly contaminated areas are presented.

704 Eckholm, Erik P. **The Picture of Health** New York: W. W. Norton, 1977, 256p. ****O**

This book addresses the environmental sources of disease such as diet, sanitation, and pollution. Topics of chapters include: undernutrition; the affluent diet; cancer; tobacco; air pollution; overpopulation; and better health care. The author stresses that although most afflictions are more typical of industrialized countries, the underdeveloped nations may suffer from the same maladies in the not-too-distant future. The notes are extensive. The list of suggested readings points toward other good sources for the environmental health hazards of disease.

705 Fast, Julius. **Weather Language**. New York: Wyden Books, 1974, 242p.

The author of *Body Language* has produced a book that describes how climate affects body and mind. Included in this discussion are sections that outline the responses of the body and mind to pollution. Pollutants are explained in the chapters that detail the effects of electricity in the air and debate if cancer is caused, in part, by weather. The reference list is an excellent source of additional material on climate, weather, and health. This is an interesting book that is only marginally relevant. The lack of an index may make finding useful information difficult.

706 Fishman, Jack, and Robert Kalish. "Smoke Gets in Your Eyes: Effects of Ozone on Human Health," in Fishman, Jack, and Robert Kalish, **Global Alert, The Ozone Pollution Crisis**. New York: Plenum Press, 1990, p. 195-225.

This chapter in a book on global ozone pollution specifically addresses the health problems of air pollution. The authors begin with a history of bad air by summarizing pollution in Donoroa, Pennsylvania, in October 1948 and in London, England, in 1952. The discussion continues with a look at the "new pollution" of the 1960s, or increased ozone concentrations near the ground, characteristic of Los Angeles. The chapter concludes with a look at safe levels of pollutants determined by the federal government.

707 Gentry, John T., Elizabeth Parkhurst, and George V. Jr. "An Epidemiological Study of Congenital Malformations," in Bresler, Jack B. (ed.), **Human Ecology, Collected Readings**. Reading, MA: Addison-Wesley, 1966, p. 223-231.

The study found a correlation between the distribution of unusually large numbers of cleft palate patients and radioactive rocks in the Hudson Valley and Adirondack Mountains of New York State. Socioeconomic class also was found to be a factor. Although this study was concerned with radioactive elements in the natural environment, results are convincing enough to warrant concern in areas where toxic waste disposal raises the level of radioactive substances in the subsurface. The article is recommended to anyone interested in the subject. Tables are numerous.

708 Godnig, Edward G., and John S. Hacunda. **Computers and Visual Stress**. Charleston, RI: Seacoast Information Services, 1990, 108p.

This book, coauthored by an optometrist, attempts to teach us how the computer in our workplace may contribute to visual stress. Symptoms include: tired eyes; blurred vision beyond close range; headaches; fatigue; double vision; and lack of concentration. The book contains six chapters: "A High Tech Health Concern"; "Visual Stress Symptoms"; "The VDT Working Environment and Visual Stress"; "Visual Training Techniques"; and "VDT Health Risks and the Individual." The glossary, list of information sources and help, and sample questionnaire to discuss with your eye care specialist are useful. Some illustrations accompany the text.

709 Gofman, John W., and Egan O'Connor. **X-Rays: Health Effects of Common Exams**. San Francisco: Sierra Club Books, 1985, 439p.

This is a practical guide that outlines the high-risk and low-risk medical and dental exams. The book begins with an overview of x-rays and the misconceptions that surround them. It continues with a chapter describing cancer as a delayed effect of x-ray exposure. The bulk of the volume, however, is a series of tables that list the average dosage and risk for certain types of tests by age. This is an essential reference for people worried about exposure to x-rays.

710 Gough, Michael. **Dioxin, Agent Orange: The Facts**. New York: Plenum Press, 1986, 290p. ****F**

This treatment of the hazards of dioxin, also known as Agent Orange, is unbiased and scholarly. The author, an expert on microbiology, is the director of the Risk Science Institute in Washington, D.C., and was formerly in the Division of Health and Life Sciences of the Congressional Office of Technology. The book addresses topics including the carcinogenic properties of dioxin, the short-and long-term effects of exposure to the substance, the possible link to birth defects, and, the subject most in the news, the cause of death of some Vietnam veterans.

711 Graves, Philip E., and Ronald J. Krumm. **Health and Air Quality: Evaluating the Effects of Policy**. Washington, DC: American Enterprise Institute for Public Policy Research, 1981, 156p.

The second part of this volume is the most relevant. In it, the authors emphasize the effects of single pollutants and of the interactions of many pollutants on human health. The approach is economic: What is the cost of the damage done to human health? The first part of the book concentrates on the issues surrounding government regulation of air pollution.

712 Greenberg, Michael R. (ed.). **Public Health and the Environment: The United States Experience**. New York: Guilford Press, 1987, 395p. ****O**

The volume contains a number of papers written by experts in the fields of health science and environmental policy. The major environmental problems and their health effects are reviewed. Most of the papers are clear and concise, some can be quite turgid. This is a good source for background information necessary to make an educated decision about environmental policies.

713 Hallenbeck, W. H., and K. M. Cunningham-Burns. **Pesticides and Human Health**. New York: Springer-Verlag, 1985, 176p.

The book is a list of pesticides that was prepared as a reference for medical and public health professionals, toxicologists, environmentalists, consumer advocates, lawyers, regulators, industrial hygienists, and users of pesticides. Under each entry the following characteristics appear: the use of the pesticide; acute exposure effects; chronic exposure effects; suspected effects; toxicology; active ingredients; and additional information. The appendices contain a glossary, a bibliography, and compilations of canceled and suspended pesticides. The alphabetical organization of the book makes it easy to use. This is an invaluable resource.

714 Hammond, E. Cuyler, "Smoking and Death Rates—A Riddle in Cause and Effect," in Bresler, Jack B. (ed.), **Human Ecology, Collected Readings**. Reading, MA: Addison-Wesley, 1966, p. 414-436.

This article was given originally as a Sigma Xi National Lecture in 1957. The author begins with a discussion of the pitfalls of scientific studies carried out on humans and continues with a description of two types of investigations that have proven to be more fruitful than others: retrospective and prospective studies of the association between smoking and lung cancer. The discussion illustrates well the reasoning behind the interpretation of ambiguous results. The figures are informative.

715 Kane, Dorothy Noyes. **Environmental Hazards to Young Children**. Phoenix, AZ: Oryx Press, 1985, 246p. ****O**

This is a thorough guide to the environmental hazards that young children face. The book is divided into two parts: 1) energy hazards and 2) materials hazards. Topics are: traffic; fires and burns; falls; noise; radiation; submersion; toxic ingestions; metals and fibers; gases and particulates. Although the volume encompasses more than direct threats from pollution, it is a valuable reference. Many figures accompany the text. Bibliographies are at the end of each chapter.

716 Langer, William L. "The Black Death," in *Scientific American*, **Man and the Ecosphere**. San Francisco: W. H. Freeman, 1971, p. 32-37.

This is an article that was published originally in *Scientific American* in February 1964. The author dramatizes that not only do humans affect the environment but the environment also touches humans. The plague of fourteenth-century Europe, in which nearly a quarter of the population perished, is described to illustrate this point. The plague was transmitted by fleas carried by rats and other rodents that proliferated in the unsanitary conditions for most European cities at that time. This is a good account of the social effects of the plagues that swept Europe in the 1300s.

717 Legator, M., B. Harper, and M. Scott. **The Health Detective's Handbook: A Guide to the Investigation of Environmental Health Hazards by Nonprofessionals**. Baltimore, MD: Johns Hopkins University Press, 1985, 300p. ****O**

This is a thorough guide to the health effects of toxic substances. The emphasis is on chemicals that may be found around the local community such as asbestos, pesticides, and herbicides. As the title suggests, the book is written for the general reader who is determined to discover possible environmental problems in the neighborhood. This is an excellent manual for the citizen who wishes to be involved. It is comprehensive and easy to use.

718 Lillie, David W. **Our Radiant World**. Ames, IA: Iowa University Press, 1986, 226p.

The author describes the health hazards and benefits of radiation and the issues surrounding radiation in clear, concise terms. Issues include nuclear power, microwave stations, radiation in medicine, and high-voltage power lines. The text begins with a straightforward and nonmathematical treatment of particulate and electromagnetic radiation. The chapters on the effects of radiation and radiation in our daily lives are the most informative. The book is suitable for anyone interested in the subject. An excellent bibliography and good index conclude the volume.

719 Loomis, W. F. "Rickets," in *Scientific American*, **Cities: Their Origin, Growth and Human Impact**. San Francisco: W. H. Freeman, 1973, p. 112-122.

Rickets was thought to be primarily from a lack of vitamin D, but is in fact due to a lack of sunlight. As such, it was one of the earliest diseases related to air pollution in the smoky cities of the late nineteenth century. The author traces its development from 1650, the time of introduction of soft coal to Europe, through the Industrial Revolution, to the present when the disease is largely eradicated. This is an interesting account of the identification of the source and the cure for an environmentally related disease.

720 McDermott, Walsh. "Air Pollution and Public Health," in *Scientific American*, **Man and the Ecosphere**. San Francisco: W. H. Freeman, 1971, p. 137-145.

This article first appeared in *Scientific American* in October 1961 and was one of the earliest to try to link air pollution to disease. Unfortunately, much of the evidence was not yet available in 1961. The author attempts to assess the consequences of long-term exposure to low levels of pollutants, such as carbon monoxide, that are known to be lethal at high concentrations. An interesting aside is that the hazardous level for carbon monoxide is cited as 150 parts per million. Recent studies have shown health problems from as little as 30 parts per million. Thermal inversion, responsible for smog in Los Angeles, is explained. The figures are excellent.

721 Panati, Charles, and Michael Hudson. **The Silent Intruder**. Boston: Houghton Mifflin, 1981, 208p.

This book addresses the effects of radiation and how to survive them. Most sources of radiation are discussed: ultraviolet rays from the sun; x-rays used in medical analysis; microwaves; and high-voltage power lines. Nuclear power plants also are described. Practical suggestions are given to avoid exposure. These include: limiting time in the sun; reducing the number of dental and medical x-rays an individual has taken; and not living close to electric power lines. The appendices contain useful information such as cancer risks from x-rays; radiation doses to a fetus from diagnostic x-ray tests; and average doses for diagnostic x-ray examinations.

722 Peters, George A., and Barbara J. Peters. **Sourcebook on Asbestos Diseases: Medical, Legal and Engineering Aspects**. New York: Garland STPM Press, 1980, 454p. ****F**

This reference for asbestos and its health effects contains eight parts. The first is a summary of types, applications, trade names, and producers of asbestos. The second outlines the illness caused by exposure to the substance. The third describes controls and protection in the form of equipment and clothing. The fourth and fifth parts concern the legal ramifications of asbestos-related injury or illness. The sixth is a questionnaire and the last two parts are bibliographies. This is a turgid volume that is recommended only to those who need legal or medical information about asbestos.

723 Scalet, Elizabeth A. **VDT Health and Safety, Issues and Solutions**. Lawrence, KS: Ergosyst, 1987, 143p.

This is an excellent introduction to video display terminals (VDTs) and their health effects. Chapters include: "VDTs and Radiation"; "Eyes and Vision"; "Strain and Injury"; "Stress"; and "Pregnancy and Reproduction." Each chapter begins with a list of major concerns and continues with a detailed discussion of the relevant information. Research findings are summarized clearly and concisely. A list of recommended readings concludes each chapter. Some illustrations, a glossary, a good index, and a thorough bibliography augment the text.

724 Skinner, H. Catherine W., Malcolm Ross, and Clifford Frondel. **Asbestos and Other Fibrous Materials**. Oxford, England: Oxford University Press, 1988, 204p.

This is an introduction to fibrous inorganic materials and a summary of their health and biological effects. Health effects are divided into the respiratory system, the pleura, and the lymphatic system. Appendices include lists of all natural and synthetic fibrous materials. The glossary of terms is very useful. The volume is written from the perspective of mineralogy.

725 Skjei, Eric, and M. Donald Whorton. **Of Mice and Molecules: Technology and Human Survival**. New York: Dial Press, 1983, 347p. ****O**

This is a comprehensive account of the health dangers created by the products of our modern society. The authors have attempted, and succeeded, to be alarming without sensationalizing the issues. Asbestos, PCBs, lead, benzene, smog, mercury, Agent Orange, paraquat, hazardous wastes, and radiation are discussed. Numerous figures illustrate how the body absorbs and responds to the toxins. Two appendices give information on emergency assistance and state agencies. Five books are suggested for further reading. This is a worthwhile study recommended to everyone.

726 Stellman, Jeanne, and Mary Sue Henifin. **Office Work Can Be Dangerous to Your Health**. New York: Pantheon Books, 1983, 239p.

This is an excellent guide to the hazards inherent in your daily work. The authors cover the full range of topics, including video display terminals, inappropriate office furniture, supplies and equipment, lighting, fire safety, and indoor air pollution. Appendices list resource organizations, describe new technology in the American office, and present an office health and safety survey. Some illustrations accompany the text.

727 Thomson, William A. R. **A Change of Air: Climate and Health**. New York: Charles Scribner's Sons, 1979, 225p.

A discussion of the effects of climate on health, this book is included to illustrate the dynamics of changing climates and their consequences. The author describes heat, cold, wind, high altitude, sunshine, and sea breezes. He also examines the relief of illness such as rheumatism, migraines, mental health, cancer, and lung disease by certain climates. The chapter on human-made climates is particularly worthwhile. The bibliography is short but provides the interested reader with several additional sources.

728 Tschirley, Fred H. "Dioxin," in *Scientific American*, v. 254, n. 2, February 1986, p. 29-36.

The author argues that evidence for serious long-term effects of dioxin on humans is lacking. He concedes, however, that it is toxic to certain animals. The article begins with a history of the family of dioxin compounds and continues with a summary of its degradation processes. Toxicity of dioxin forms the bulk of the article. This is an interesting article written from a different perspective than most.

729 Upton, Arthur C. "The Biological Effects of Low-Level Ionizing Radiation," in *Scientific American*, v. 246, n. 3, February 1982, p. 41-49.

Written by the chairman of the Institute of Environmental Medicine at New York University School of Medicine, the article discusses the effects of low-level radiation from both natural and artificial sources, including cosmic and terrestrial radiation, radioactive isotopes within our bodies, global fallout, and medical analyses. The damage to DNA and to chromosomes is described. Carcinogenic effects are the most typical of low-level radiation exposure. A figure that depicts the perception of risk of sources of death by three groups—the League of Women Voters, college students, and members of business and professional clubs—is fascinating.

730 Watanabe, Tsutomu. "Infectious Drug Resistance," in *Scientific American*, **Man and the Ecosphere**. San Francisco: W. H. Freeman, 1971, p. 189-197.

The article was published originally in *Scientific American* in December 1967. The spread of infectious disease is not a topic most people believe is related to man's environmental problems. The spread of disease, however, is related directly to the density and mobility of populations and the general level of public health. Overpopulation and malnutrition are serious environmental problems in areas where sanitation is not the best. In addition, drug-resistant bacteria are proliferating. This article illustrates these points dramatically. The last few paragraphs are fairly technical, but otherwise the article is suitable for everyone.

731 Whelan, Dr. Elizabeth. **Toxic Terror**. Ottawa, IL: Jameson Books, 1985, 348p. ****O**

The book attempts to tell the truth about the relationship between technology and cancer. The author argues that most chemicals proclaimed as carcinogenic by alarmists are, in fact, harmless and can be, in some cases, extremely beneficial. She further stresses that the "toxic terrorists" are damaging health and the environment. One example is the return of malaria (which had been largely eradicated) due to disappearance of effective drugs now deemed unsafe. Pesticides, PCBs, PBBs, dioxin, formaldehyde, Love Canal, acid rain, water and air pollution, and nuclear energy are discussed. The appendix lists cancer statistics. A provocative and engrossing work, this book is recommended to anyone interested in an alternative perspective.

732 Ziff, Sam. **Silver Dental Fillings: The Toxic Time Bomb**. Santa Fe, NM: Aurora Press, 1984, 197p.

The author summarizes the available information on mercury toxicity from dental amalgams. Silver fillings can be as much as 40 percent mercury. Topics include: mercury in modern medicine; arguments for and against the use of mercury; the transport of mercury in the body; the residence time of mercury in the body; and the physical and psychological effects of mercury ingestion. A section that outlines alternatives to silver fillings concludes the book. References are at the end of each chapter. This is a reasonably unbiased treatment of a subject that receives more attention each day.

16

Solutions to Environmental Problems

The severity of the world's environmental problems is beginning to become apparent to people from all walks of life: industrialists; politicians; business executives; journalists; government officials; developers; religious figures; activists; homemakers; college students; and school children. Most of these individuals now believe that long-term survival of the planet and human society may depend on working closely with one another. Corporate goals should not be pursued at the expense of the environment. Recycling, once dismissed by industry, for example, can be profitable. Similarly, environmental concerns should consider the economic conditions of a region. In the rainforests, development and ranching are ways to reduce the poverty that has become a way of life. Offering alternatives to clear-cutting that are economically viable will slow the destruction much more rapidly than simply banning slash-and-burn practices.

In addition to priorities changing within business, political, and government circles, concerns are shifting on the local scale—in communities, schools, offices, and homes. Individuals are anxious to do things that will help and to avoid actions that will harm our already fragile and damaged ecosystems. These may range from informing oneself and getting politically involved to installing low-flow shower heads and riding bicycles instead of driving cars. The decisions we make, no matter how small or how large, will affect the world in which we live.

The works in this chapter offer solutions to environmental problems. Some are philosophical and political treatments of complex issues facing governments and industry. Most, however, are manuals that summarize the global environmental condition and list practical suggestions to help improve it. Excellent guides to what the individual can do include: *Our Earth, Ourselves* by Ruth Caplan and the Staff of Environmental Action; *State of the Government* by the Conservation Foundation; *The Global Ecology Handbook* by Walter H. Corson and the Global Tomorrow Coalition; *50 Simple Things You Can Do to Save the Earth* by the Earthworks Group; *The Green Consumer* by John Elkington, Julia Hailes, and Joel Makower; *Save Our Planet* by Diane MacEachern; *The Green Lifestyle Handbook*, *1001 Ways You Can Heal the Earth* by Jeremy Rifkin; *Blueprint for a Green Planet* by John Seymour and Herbert Girardet; and *Saving the Earth: A Citizen's Guide to Environmental Action* by Will Steger and Jon Bowermaster. A person who wishes to examine just one book to learn about the environment would do well to read one of these. A good factual reference is *The Earth Report: The Essential Guide to Global Ecological Issues* by Edward Goldsmith and Nicholas Hildyard, which contains six essays on environmental problems and an encyclopedia of over 400 keywords, concepts, and statistics. Other topics are: the rainforest (note Scott Lewis); successful community projects (note John J. Berger, Michael Corbett); living with toxic substances (note Edward J. Bergin, Gary Cohen and John O'Connor, Nicholas Freudenberg); renewable energy resources and conserving energy (note Harry Braun, Joseph J. DiCerto, Peter Norback and Craig Norback); product safety (note Stephen Brobeck and Anne C. Avery); garbage (note Marian R. Chertow); sustainable human development (note William C. Clark, Jim MacNeill, Francis R. Thibodeau and Hermann H. Field); efforts to combat pollution since the 1960s (note Barry Commoner); environmental choices (note Kathleen Courrier, Christopher Plant and Judith Plant); deep ecology and the green movement (note Bill Devall, Brian Tokar); environmental clean-up (note Gregg Easterbrook, Jeremy Main); policies for the decades to come (note William R. Ewald, Jr., Jonathon Weiner); monkey-wrenching (note David Freeman and Dave Haywood); strategies for manufacturing (note Robert A. Frosch and Nicholas E. Gallopoulos); agriculture (note Wes Jackson); "green" businesses (note Melanie Menagh); environmental vacations (note Stephanie Ocko); diet (note John Robbins); health (note Harold J. Taub, Carol Venolia); and environmental policy (note Norman J. Vig and Michael E. Kraft, Jonathon Weiner, William R. Ewald, Jr.). Numerous environmental books are published each month and most fall into this category. The interested individual is sure to find additional helpful material in the local library or bookstore.

733 Berger, John J. **Restoring the Earth: How Americans Are Working to Renew Our Damaged Environment**. New York: Alfred A. Knopf, 1985, 241p.

This book is an optimistic look at individuals and groups who have struggled to restore our damaged planet. People whose tales are told include a California pharmacist who planted 10,000 trees and saved a redwood forest, a Wisconsin architect who redesigned a town prone to flooding into a floodproof solar city, and a Cape Cod plumber who turned a ditch into a trout stream. The author also lists several proposals to make our cities more ecologically sound.

734 Bergin, Edward J., with Ronal E. Grandon. **How to Survive in Your Toxic Environment**. New York: Avon Books, 1984, 505p.

The authors describe the "chemical poisoning of America" by the large chemical manufacturers and illustrate how to survive this onslaught. The discussion begins with myths surrounding the chemical industry such as, "It costs too much to clean it up" or "They wouldn't sell it if it wasn't safe." The authors continue with essays on a variety of topics, including regulation and the inadequacy of labeling. The bulk of the book is a practical manual for identifying hazards both in the workplace and at home and for protecting your rights as a citizen. The appendices are very useful and cover such subjects as a guide to occupational/environmental diseases, hazardous waste information sources, and toxic chemical/hazardous waste consulting firms. Although the work is biased, it is a useful reference volume and has many excellent points and suggestions.

735 Braun, Harry. **The Phoenix Project: An Energy Transition to Renewable Resources**. Salt Lake City, UT: Publishers Press, 1990, 272p.

The book identifies the solar technologies that are economic and practical alternatives to fossil and nuclear fuels. The emphasis is on renewable solar-hydrogen resources, which possess the ability to reduce greenhouse gases in the atmosphere, acid rain, and chlorofluorocarbons (CFCs) without causing other types of pollution. The author outlines the steps required to change to a "solar-hydrogen based economy" and argues the change will be beneficial to both the United States and the former U.S.S.R. Included in the book are reviews of global environmental problems, possible solutions, the relationships between energy and the environment, the use of hydrogen as a fuel, viable solar technologies, and hydrogen legislation. The text is accompanied by numerous illustrations, a reference list, and an index.

736 Brobeck, Stephen, and Avery, Anne C. **The Product Safety Book**. New York: E. P. Dutton, 1983, 440p.

The book is an alphabetical listing of products. Under each entry, the hazards of and precautions for using the product are outlined. Also listed in some cases is a chronology of mishaps and a definition. The encyclopedia format is convenient and easy to use. Cross-referencing is extensive. Products are as diverse as hot dogs, mini-bikes, sardines, power saws, and estrogen. The catalogue begins with a guide on its use, which includes a list of acronyms. This is a great reference put out by the Consumer Federation of America.

737 Caplan, Ruth, Executive Director, and the Staff of Environmental Action. **Our Earth, Ourselves**. New York: Bantam Books, 1990, 340p. ****O**

Written by the people who founded Earth Day, this book is comprehensive and practical, offering suggestions such as avoiding solid air fresheners to reduce chemical pollution in the home, walking to decrease the amount of emissions from car exhaust in the air, and boycotting products in nonrecyclable packaging to alleviate the landfill problem. Nine chapters discuss global warming, ozone depletion, air pollution, toxic waste, landfills, the nuclear industry, rainforests, overpopulation, and water pollution. Each chapter begins with an overview of the problem and ends with suggestions for what can be done in the home, the community, and the nation. The last chapter is a guide to becoming an environmental activist. A list of organizations and recommended reading follow the chapter. The introduction is by noted folksinger and activist Pete Seeger.

738 Center for Renewable Resources. **Shining Examples**. Washington, DC: Center for Renewable Resources, 1980, 210p.

This is a catalog of model projects throughout the United States that use renewable resources. The purpose is to describe a cross-section of projects to illustrate that the transition to alternative resources is underway and, at least locally, successful. The authors also wish to put the energy crisis in perspective. The projects are classified into eight groups: agricultural; community; commercial and industrial; education and outreach; financing; government; housing and public buildings; and utility. This is an extremely informative and useful book.

739 Chertow, Marian R. **Garbage Solutions: A Public Official's Guide to Recycling and Alternative Solid Waste Management Technologies**. National Resource Recovery Association, U.S. Association of Mayors, 1989, 74p.

The author, director of the Program on Solid Waste and the Environment in the Institution for Social and Policy Studies at Yale University, looks realistically at the management of solid waste in a community. She concludes that reducing the amount of garbage produced, recycling, and composting can have a significant effect on the waste management program. Written in non-technical language, the book is a good overview of municipal solid waste management.

740 Clark, William C. "Managing Planet Earth," in *Scientific American*, v. 261, n. 9, September 1989, p. 46-54. **S

The author, a senior research associate at the Kennedy School of Government at Harvard University, analyzes the prospects for "sustainable human development on a planet with finite resources and a fragile environment." He also describes the type of planet that may develop once the goals of sustainable development are met. Topics discussed include the climate, chemical cycles of nutrients, the hydrological cycle, and biological diversity. "Adaptive planetary management" is stressed. The figures are informative and complement the text. This is an excellent overview article recommended to all.

741 Cohen, Gary, and John O'Connor (eds.). **Fighting Toxics: A Manual for Protecting Your Family, Community and Workplace**. Washington, DC: Island Press, 1990, 346p.

Prepared by the National Toxics Campaign, this book is a guide for citizens to help reduce pollution in their communities. The editors have compiled effective strategies used by people all over the United States to combat toxic substances. Suggestions are given for getting information, inspecting the neighborhood, working with the media, achieving legal recourse, and initiating corporate campaigns. Also included are descriptions of federal statutes, the magnitude of the toxics crisis, and local campaigns. A list of resources concludes the book. This is a pragmatic book that is indispensable to anyone interested in being actively involved in limiting toxics in the community.

742 Commoner, Barry. **Making Peace with the Planet**. New York: Pantheon Books, 1990, 292p.

The noted environmental philosopher and scientist has written a book that reviews the efforts that have been made to combat pollution since the "eco-revolution" of the 1960s and 1970s. His conclusion is that the reality has fallen far short of the expectation. Consumer patterns were not changed. The government miscalculated on several occasions. Private industry resisted. The author argues that pollution is an "incurable disease" that can only be prevented by redesigning the way Americans manufacture goods. Specific topics include: garbage incineration; dioxin; Bhopal; fossil fuel emissions; the greenhouse effect; and "acceptable risk." This is a fascinating book that is accessible to everyone. Extensive notes and an index accompany the text.

743 Commoner, Barry. "A Reporter At Large: The Environment," in *The New Yorker*, June 15, 1987, p. 46.

The author, the director of the Center for the Biology of Natural Systems at Queens College, has written an eloquent explanation for why the current environmental predicament will not disappear unless modern manufacturing technologies are changed. His argument is that cleaning up is not sufficient: the pollutant must be eliminated. This is an interesting, if somewhat long-winded, article.

744 The Conservation Foundation. **State of the Government: A View Toward the Nineties**. Ann Arbor, MI: Books on Demand. 1987, 664p. ****O**

This report by the nonprofit research organization, the Conservation Foundation, analyzes the state of the environment and offers suggestions to improve it. The emphasis is on pesticides and the production of food, the protection of endangered species, the disposal of hazardous waste, and air pollution. The discussion is lucid and informed.

745 Corbett, Michael. **A Better Place to Live: New Designs for Tomorrow's Communities**. Emmaus, PA: Rodale Press, 1982, 164p.

The author believes that the houses we construct are too big and ecologically unsound. He urges that smaller houses be built that are more in harmony with their surroundings. If the small space is cleverly designed, most people will not notice its size. With smaller homes and communities, the author feels that human growth and fulfillment are easier to attain.

746 Corson, Walter H., and The Global Tomorrow Coalition. **The Global Ecology Handbook**. Boston: Beacon Press, 1990, 414p. ****O**

This guide to what the individual can do to help the environment is the companion volume to the Public Broadcasting Service's television series "Race to Save the Planet." Practical tips include: assessing the environmental quality of your home and office; conserving water; reducing garbage; changing your gardening habits to protect animal and plant life; establishing a recycling program; and getting involved. The handbook is organized by environmental problem, such as tropical rainforests, population growth, global warming, hazardous waste, and fresh water, which allows summaries of key concepts and important issues to be included with the suggestions for action. The appendix is a list of organizations mentioned in the text. Abundantly illustrated, this is an excellent reference for the environmentally concerned family.

747 Courrier, Kathleen (ed.). **Life After '80: Environmental Choices We Can Live With**. Andover, MA: Brick House, 1980, 280p.

The volume is a collection of essays by 35 authors. Some of the contributors are Jacques Cousteau, Barry Commoner, Anne and Paul Ehrlich, Michael Brown, Stephen Schneider, Amory Lovins, and Denis Hayes. Essays are grouped into six parts: 1) the elements, which discusses the sea, rivers, forests, soil, deserts, climates, and the atmosphere; 2) inhabitants; 3) legacies, which focuses on waste and cancer; 4) eco-logic; 5) renewable energy; and 6) coming to terms. The format allows the presentation of opposing viewpoints. This is an interesting and provocative book.

748 Devall, Bill. **Simple in Means, Rich in Ends: Practicing Deep Ecology**. Salt Lake City, UT: Peregrine Smith Books, 1988, 224p.

This is a guide for living without harming the Earth. The author suggests how to discover one's "ecological self," what to do for political action, and how to adapt one's career to embrace the principles of deep ecology. Monkeywrenching, ecotage, and green politics are explained. The appendices contain a list of books for further reading and a bibliography. The writing is straightforward and easy to follow. This work is suitable for the general reader who wishes to learn more about commitment to the environment.

749 DiCerto, Joseph J. **The Electric Wishing Well, the Solution to the Energy Crisis**. New York: Macmillan, 1976, 317p.

The author, speaking from the perspective of the energy crisis of the early 1970s, remains optimistic for the future. He describes the dependence of modern society on fossil fuels but outlines numerous alternatives such as hydroelectric power, solar energy, and geothermal energy. New technologies will be manufactured to take advantage of all energy possibilities. Examples include the bombardment of dust-sized pieces of deuterium by 12 lasers and the creation of dams and windmills to harness ocean tides. The author makes it all seem so possible. A bibliography and index conclude the volume.

750 Earthworks Group. **50 Simple Things You Can Do to Save the Earth**. Berkeley, CA: Earthworks Press, 1989, 96p. ****O**

This book sat atop the *New York Times* nonfiction bestseller list for many weeks. Each page is devoted to a suggestion for improving the state of the global environment. These include: turning off the tap while brushing your teeth; not pulling the vapor catcher back on the gas pump nozzles when filling your car; recycling aluminum cans; planting a tree; and using plants indigenous to your area in your garden. A section called "What's Happening" starts the book. It contains quotes from environmental groups explaining problems such as the greenhouse effect, ozone, CFCs, acid rain, and hazardous waste. This is a very useful book that can help individuals feel they are contributing to a solution.

751 Easterbrook, Gregg. "Cleaning Up," in *Newsweek*, July 24, 1989, p. 27.

The article discusses a proposal by President Bush for an "emission-trading system" to fight acid rain. Those companies that adhere to federal guidelines are rewarded monetarily. Those that do not must buy "emissions-reductions credits." In the system, engineers will decide which factories can meet strict standards. The regulator's role will be reduced. In addition, a profit motive would be introduced for complying with regulations and for designing better control equipment.

752 Elkington, John, Julia Hailes, and Joel Makower. **The Green Consumer**. New York: Penguin Books, 1988, 342p. ****O**

This is an indispensable reference for the consumers concerned about how their actions affect the environment. The handy guide is divided into three parts: an introduction that explains what "green" is; a shopping list of environmentally sound and hazardous items; and a manual on how to get involved. Automobiles, home furnishings, personal care products, and garden supplies are analyzed. The index makes this a very convenient book to use.

753 Ewald, William R., Jr. (ed.). **Environment and Change: The Next Fifty Years**. Bloomington, IN: Indiana University Press, 1968, 397p.

A collection of papers that were presented originally at a conference of the American Institute of Planners in Washington, D.C., in 1967, this volume attempts to predict all aspects of environmental change in the next few decades. Topics include: the future of man; the arts in modern society; the role of technology in creating the environment; the culture of urban America; individuals in a collective society; and technology and the underdeveloped world. The papers are especially interesting from the perspective of the 1990s. This is a thoughtful, philosophical collection that is a bit outdated.

754 Foreman, David, and Dave Haywood. **Ecodefense: A Field Guide to Monkeywrenching, 2d Edition**. Tucson, AZ: Earth First! Books, 1987, 312p.

Written by the two founding members of Earth First!, a radical environmental organization whose motto is "No compromise in the defense of Mother Earth," this book outlines the often illegal methods of "monkey-wrenching" and "ecotage" (ecological sabotage). Techniques such as tree-spiking, in which trees are laced with metal spikes such that damage is done to equipment during logging, billboard removal, and fence cutting are summarized. The handbook is a good overview of what Earth First! is all about.

755 Freudenberg, Nicholas. **Not in Our Backyards: Community Action for Health and the Environment**. New York: Monthly Review Press, 1984, 304p.

The book is a manual on how to fight the environmental hazards produced by toxic wastes in our communities. Ten chapters discuss citizen rights and actions, first by giving an overview of the subject and then by outlining a case history of a community's struggle to remove the environmental threat. Love Canal is one example. Topics include: environmental health research; federal regulations; community organization; forming an effective organization; getting information; and building coalitions. References, an index, and a list of resources conclude the book.

756 Frosch, Robert A., and Nicholas E. Gallopoulos. "Strategies for Manufacturing," in *Scientific American*, v. 261, no. 9, September 1989, p. 144-152.

The authors illustrate how waste from one industry can be used as the materials in another in a form of recycling that reduces the impact on the environment. They begin by discussing an industry where the effects of waste were unknown: the conversion from rawhide to tanned leather. The process created smells and disease such that tanneries had to be moved from communities. Another example is CFCs, which were considered miracle products until they were discovered to destroy ozone. In contrast, the iron industry recycles most of its material, thereby minimizing the amount of waste. This is an interesting article accompanied by many photographs and figures that is a good introduction for the general reader.

757 Goldsmith, Edward, and Nicholas Hildyard. **The Earth Report: The Essential Guide to Global Ecological Issues**. Los Angeles: Price Stern Sloan, 1988, 240p. ****F**

This complete handbook for the general reader on environmental problems is organized into two parts. The first has six essays by noted environmental experts on a variety of topics: man and the natural order; the politics of food aid; nuclear energy after Chernobyl; man and Gaia; acid rain and forest decline; and water fit to drink. The second part is an encyclopedia containing over 400 entries of concepts, keywords, and ecological facts. Examples are: landfill; leukemia clusters; erosion; heavy metals; resettlement programs; and 2, 4-D. The volume has numerous illustrations and an index.

758 Harris, D. Mark. **Embracing the Earth: Choices for Environmentally Sound Living**. Chicago: Noble Press, 1990, 192p.

The author has written a simple and practical guide to living with less impact on the Earth. The book has four parts, each of which contains several chapters. The first, "Starting Out," explains the three R's—reduce, reuse, and recycle—and suggests how to eliminate toxics, conserve water, and save energy. Helpful hints include: buying in bulk to save packaging; stopping junk mail; using cloth instead of paper; installing adequate insulation; and washing dishes by hand as opposed to in the dishwasher. The second part of the book, "Getting More Involved," emphasizes what the individual can do to affect the community at large. Projects are: planting more trees; lobbying and writing legislators; and joining environmental organizations. The third part of the book, "Going All the Way," focuses on education and the state of the world. The fourth and final part, the directories, lists environmental organizations and members of Congress.

759 Heloise. **Hints for a Healthy Planet**. New York: Perigree Books, 1990, 160p.

The helpful columnist has put together a book containing practical suggestions to reduce waste and increase environmental awareness. The book is divided into five sections: inside the house; outside the house; in the workplace; play and entertainment; and the final word. Some hints include: putting lids on pots to allow water to boil more quickly (uses less heat); placing coffee grounds around azaleas, rhododendrons, and roses; and drinking coffee at work from your own mug to cut down on styrofoam cups. The introduction summarizes the landfill problem. Heloise practices what she preaches: her book is printed on recycled paper.

760 Hynes, H. Patricia. **Earth Right: Every Citizen's Guide**. Rocklin, CA: Prima Publishing & Communications, 1990, 236p. ****S**

The author, a professor of urban studies and planning at MIT, has put together a guide for people seeking "lucid, accessible, and practical ideas about reversing environmental destruction." The book has five chapters: 1) "Pesticides: Poisons in a Progressive Package"; 2) "Solid Waste: Treasure in Trash"; 3) "Drinking Water: Toxins on Tap"; 4) "The Ozone Layer: Earth's

Evanescent Membrane"; and 5) "Global Warming: The Greenhouse Effect." Each chapter contains a nontechnical explanation of the problem, a series of actions to take to alleviate the problem, and a list of resources available to the average citizen. This is an excellent and informative treatment.

761 Jackson, Wes. **New Roots for Agriculture**. San Francisco: Friends of the Earth, 1980, 155p.

The author is the director of the Land Institute, a nonprofit, educational research organization studying sustainable alternatives in agriculture, energy, and waste management. In this book, the practice of "till agriculture"—essentially all modern farming methods—is called into question because it is not sustainable and it destroys the soil. Planting crops that act like a natural ecosystem is advocated. The discussion is broken into four parts. The first defines the problem, the second argues a new philosophy, the third outlines solutions, and the fourth describes a vision for the future.

762 Lamb, Marjorie. **2 Minutes a Day for a Greener Planet: Quick and Simple Things Americans Can Do to Save the Earth**. San Francisco: Harper & Row, 1990, 243p.

The author states that she, like the rest of us, is busy. The suggestions in her book, therefore, are quick and easy to implement. Eighteen chapters discuss topics such as saving water, paper, packaging, and energy, cleaning without toxic substances, renovating homes, gardening, and driving. The polluting effects of the old methods and substances are contrasted with the benefits of the new. Alarming facts are highlighted in the text by a pointing finger. Suggestions are boldfaced and augmented by an illustration of a stopwatch with a map of the world on its face. The author and publisher believe what they are writing: two trees were planted for each one destroyed in the printing of the book.

763 Levin, Melvin R. "Environmental Costs and Benefits," in Woods, Barbara (ed.), **Eco-Solutions**. Cambridge, MA: Schenkman, 1972. p. 445-447.

The author argues that the dollar amount required to reduce pollution is offset by the expenditures saved on illness, health costs, and other environmentally produced problems. In addition, environmental regulation, control, and clean-up will generate thousands of jobs and substantial profits. The author specifically analyzes population, air pollution, water pollution, noise pollution, land use, solid wastes, and aesthetics. Several amusing cartoons accompany the text. The notes are extensive. Sources include both journal articles and books.

764 Lewis, Scott, with the Natural Resources Defense Council. **The Rainforest Book**. Los Angeles: Living Planet Press, 1990, 112p.

This concise guide to the world's rainforests is indispensable reading for those who consider themselves environmentalists. The author begins by describing the diversity of plant and animal life in the rainforest, continues by outlining the threats to survival of the rainforest, and concludes by listing the ways a member of a technologically advanced society can help stop the destruction of the rainforest. Practical tips include: avoiding woods harvested from the rainforest; purchasing products made from renewable rainforest resources; visiting American rainforests in Hawaii, Puerto Rico, the U.S. Virgin Islands, and American Samoa; and educating your children. The introduction is by Robert Redford. The last chapter tells how to get more information. The book loses a little credibility when it delves into other environmental issues (talking about carpooling, for example, to reduce the amount of fossil fuels burned, which is an important issue but not directly related to rainforest destruction).

765 MacEachern, Diane. **Save Our Planet: 750 Everyday Ways You Can Help Clean the Earth**. New York: Dell, 1990, 210p. ****O**

This book begins with an introduction that describes the environmental problems facing humankind today. Subsequent chapters outline things that individuals can do to help stop pollution. The chapters are organized by places where people lead their lives, including home, garden, garage, supermarket, school, and office. The last chapter is a discussion of vacations. The appendix lists references and organizations that address environmental concerns. The author has written a useful book that everyone will enjoy.

766 MacNeill, Jim. "Strategies for Sustainable Economic Development," in *Scientific American*, v. 261, n. 9, September 1989, p. 155-165.

The author is secretary-general of the World Commission on Environment and Development. In this article, he argues that economic growth can occur with minimal impact on the environment despite the current trend for countries to deplete their resources faster than they can be replaced. Population growth rates must decrease, whereas production rates must increase to attain sustainable development. Demand also must be reduced. This is an intelligent, straightforward article. Of special interest is a discussion on eliminating poverty in a developing country.

767 Main, Jeremy. "Here Comes the Big New Cleanup," in *Fortune*, November 21, 1988, p. 102.

The major environmental issues of the coming decade are discussed. The author advocates market-oriented incentives for the industries to encourage the cleanup of wastes. He believes that additional regulations imposed by Congress will only make the process more expensive and litigation an acceptable alternative.

768 Menagh, Melanie. "The Business of Being Green," in *Omni*, v. 13, no. 9, June 1991, p. 42-48.

This is an article about businesses that invest in "green" companies and "green" products. One example is Seventh Generation, Inc., a mail-order catalogue business in Burlington, Vermont, that sells only environmentally friendly products. The author introduces the concept of a "double bottom line"—one financial, the other environmental—and discusses abandoning the practice of "planned obsolescence."

769 Naar, Jon. **Design for a Livable Planet**. New York: Harper & Row, 1990, 338p.

The citizen frustrated by the state of the world would do well to read this book full of practical suggestions on how to clean up the environment. The author has compiled a scientifically accurate and easily read action guide with a positive outlook. Also included are overviews of most of the important environmental problems, including toxic waste, landfills, water and air pollution, acid rain, deforestation, global warming, radiation, and energy. Additional chapters on environmental law, eco-action, and the power of green are highly informative. Enthusiastically recommended, this is one of the best overviews around.

770 Norback, Peter, and Craig Norback. **Consumer's Energy Handbook**. New York: Van Nostrand Reinhold, 1981, 362p.

This is an informative and pragmatic guide to conserving energy in today's society. Tips include: adding insulation to your home; keeping your car tuned; and installing efficient appliances. Written before the current state of awareness of the environment, the book is timely and relevant.

771 Ocko, Stephanie. **Environmental Vacations: Volunteer Projects to Save the Planet**. Santa Fe, NM: John Muir Publications, 1990. 235p.

This timely book was written to suggest ways in which vacationers could aid scientific investigations of global environmental problems. The author describes possible trips, including monitoring an active volcano in Costa Rica, interviewing government officials involved in the AIDS epidemic in Africa, and building an experimental city in Arizona. The organizations that match volunteers and scientists are listed. The book also contains practical advice such as necessary medicines and clothing. The last section catalogues endangered species by country. This has some unique recommendations for spending your holiday time.

772 Plant, Christopher, and Judith Plant. **Turtle Talk: Voices for a Sustainable Future**. Philadelphia: New Society Publishers, 1990, 133p.

This is a collection of interviews with artists, activists, and organizers about the future of the Earth. Each entry tells about someone doing something about the environment. Topics include: deep ecology; endangered species; and bioregionalism. This is an interesting and optimistic book that may be a little too "New Age" for some readers.

773 Rifkin, Jeremy. **The Green Lifestyle Handbook: 1001 Ways You Can Heal the Earth**. New York: Henry Holt, 1990, 198p. ****O**

This authoritative and informative book offers hundreds of suggestions for changing the ways we live to help establish a sustainable future for the Earth. The book has 21 chapters, each written by a different author, on topics ranging from household toxics to boycotts, arranged in four sections: 1) back to basics; 2) lifestyles; 3) cultivating solutions; and 4) getting organized. Three appendices are included: 1) an explanation of global warming, the greenhouse effect, and acid rain; 2) a summary of nuclear power and genetic engineering; and 3) an annotated guide to environmental books. The introduction is an excellent overview of the current, serious global pollution threat.

774 Robbins, John. **Diet for a New America**. Walpole, NH: Stillpoint, 1987, 423p.

The author, using medical, agricultural, economic, and environmental studies as evidence, argues that the United States would be a far better place and its population healthier, if Americans were to give up eating animal proteins and animal fats. Obvious benefits are mentioned such as decreases in cancer and heart disease. Also discussed are more obscure consequences: less deforestation for grazing; less water use for irrigation of ranchland; and fewer toxins in our diet by consuming lower down the food chain. The author has an interesting point of view that will convince many people to change their diets. A list of resources is included.

775 Ruckelshaus, William D. "Toward a Sustainable World," in *Scientific American*, v. 261, n. 9, September 1989, p. 166-174.

Written by a former head of the Environmental Protection Agency, this article is a serious but optimistic look at how to formulate the policies that will permit development and growth without severely impacting the environment. The task is not easy, the author argues, because environmental consequences are not immediate. In fact, some scientists suggest they may never occur. To accomplish sustainable development, three beliefs must prevail: 1) humans are a part of nature and draw their sustenance from their surroundings; 2) economic activity must account for the environmental costs of production; and 3) maintenance of the global environment depends on the sustainable development of all humans. This is a well-stated case that is easy to follow.

776 Scheffey, Andrew J. W. "Conservation Commissions in Massachusetts," in Woods, Barbara (ed.), **Eco-Solutions**. Cambridge, MA: Schenkman, 1972, p. 257-288.

The author describes conservation commissions in Massachusetts, of which the first was established in 1957. By 1971, over 80 percent of the cities and towns in the state had created commissions to oversee development for the benefit of residents and act as checks to ensure conservation. Programs and problems adopted and identified by specific commissions are discussed. The texts of the Massachusetts Conservation Act, the Conservation Fund Act, and the Massachusetts Self-Help Act are included at the end of the article.

777 Seymour, John, and Herbert Girardet. **Blueprint for a Green Planet: Your Practical Guide to Restoring the World's Environment**. New York: Prentice-Hall, 1987, 192p. ****O**

This begins with an overview of global environmental problems, such as deforestation, acid rain, and toxic waste, to illustrate dramatically the need for individual action. Because most damage comes from the consumer-oriented industrialized countries, individuals can change the situation quickly. Tips are given on altering daily activities, including eating, washing, and driving for minimizing environmental impact. The book is well illustrated. It has numerous concrete suggestions and is an invaluable reference.

778 Steger, Will, and Jon Bowermaster. **Saving the Earth: A Citizen's Guide to Environmental Action**. New York: Alfred A. Knopf, 1990, 306p. ****O**

The authors, an environmentalist who led the international Trans-Antarctica Expedition and a journalist who writes about environmental issues, have composed a lucid and graphically striking portfolio on major environmental threats and their solutions. Topics include: global warming; ozone depletion; acid rain; hazardous waste; smog; overpopulation; energy; and marine pollution. The book makes specific recommendations, identifies other sources of information, illustrates the major points with storyboards at the ends of chapters, and lists environmental organizations. This is an excellent resource that will go a long way in answering most people's questions.

779 Taub, Harold J. **Keeping Healthy in a Polluted World**. New York: Harper & Row, 1974, 246p.

This is designed as a program the individual can follow to protect his health in a polluted world. Hazards include: heavy metals in food; disease-carrying organisms in water; radioactive materials; and carcinogens. Chapter titles are "Vitamins Can't Breathe for You"; "Treated Water Is No Treat"; "There's Death in Your Garage"; "Polluted Food and a Woman's Breasts"; and "Just a Little Cancer." The suggested reading list is thorough. The book is well written and easy to follow. The course of action recommended in 1974 is still relevant.

780 Thibodeau, Francis R., and Hermann H. Field (eds.). **Sustaining Tomorrow**. Hanover, NH: University Press of New England, 1984, 186p.

This collection of 20 papers by noted experts presents solutions to today's environmental problems. The emphasis is on conservation and sustainable development. Papers are grouped in five categories: introduction to world conservation; conservation objectives; priorities for national action; priorities for international action; and conclusion. This is a refreshing alternative to the literature that dwells upon problems in the environment. The papers are informative, thought-provoking, and well written. A good bibliography and index accompany the text.

781 Tokar, Brian. **The Green Alternative**. San Pedro, CA: R. and E. Miles, 1987, 184p.

Written by a biophysicist committed to Green principles, this is an account of the Green movement in the United States. The ideas and roots of the movement are described concretely. The detailed "notes on sources" section contains numerous valuable references for global environmental problems. The work is essential reading for anyone interested in the Green movement in the United States or who is familiar with the group in Europe where they originated and receive substantial media attention.

782 Venolia, Carol, **Healing Environments**. Berkeley, CA: Celestial Arts, 1988, 224p.

The book is a New Age guide to indoor well-being. The author, an architect, describes ways to increase the harmony in your life by improving your indoor space. Some suggestions are basic and concrete. Subjects include: light; color; the thermal environment; sound and noise; indoor air quality; plants and gardens; design elements; and healing centers. A bibliography and index conclude the book.

783 Vig, Norman J., and Michael E. Kraft (eds.). **Environmental Policy in the 1990's**. Washington, DC: CQ Press, 1990, 418p.

The editors, two professors of political science, have collected 17 papers into one volume in an attempt to identify the most important developments in environmental policy since the 1960s and to determine the central issues for the 1990s. The book has five parts: 1) environmental policy and politics in transition; 2) public policy dilemmas; 3) environmental dispute resolution and policy integration; 4) toward global environmental policy; and 5) ethics, values, and the future of environmental politics. The "NIMBY syndrome," deadlocks in Congress, and the judicial system are described. Two appendices, one citing major environmental legislation and the other listing budget information for the 1980s, are included.

784 Weiner, Jonathon. **The Next One Hundred Years, Shaping the Fate of Our Living Earth**. New York: Bantam Books, 1990, 312p.

This is a summary of the state of the Earth at the close of the twentieth century and an estimate of the future of the planet. Most major environmental topics are discussed, including ozone depletion, global warming, greenhouse gases, species extinction, marine pollution, and chlorofluorocarbons (CFCs). The story is told through interviews with scientists investigating the various phenomena. One researcher is the noted geochemist, Charles Keeling, who measures carbon dioxide levels in the atmosphere. The explanations of the implications of the results of the different scientific studies are simple and greatly clarify the state of the Earth for the general reader. The author concludes that the situation for the survival of human beings is dire. Solutions, however, are possible. The necessary actions are described in the last part of the book.

785 Woods, Barbara (ed.). **Eco-Solutions**. Cambridge, MA.: Schenkman, 1972, 540p.

This book presents possible solutions to the environmental crisis by weighing the cost of the solution against the benefit it achieves. Discussions are lucid and realistic, refraining from becoming too alarmist. The approach assumes that having every river, ocean, mountain, and desert remain pristine is impossible without dire effects on the way of life in modern, industrialized nations. Each chapter addresses a different environmental problem such as population, air pollution, water pollution, etc., and contains articles written by experts in the field. This volume is recommended to people interested in the pragmatic approach to environmental problems.

Author/Title Index

Numbers cited in the index are entry numbers. Books are in italics and articles are in quotes.

Subject Index

Numbers cited in the index are entry numbers.